3D PRINTING MADE EASY FOR NEWBIES AND HOBBYISTS

A QUICK-START GUIDE TO LEARN HOW TO 3D PRINT AT HOME

NATALIE GUZMAN

© **Copyright 2021 - All rights reserved.**

The content contained within this book may not be reproduced, duplicated or transmitted without direct written permission from the author or the publisher.

Under no circumstances will any blame or legal responsibility be held against the publisher, or author, for any damages, reparation, or monetary loss due to the information contained within this book, either directly or indirectly.

Legal Notice:

This book is copyright protected. It is only for personal use. You cannot amend, distribute, sell, use, quote or paraphrase any part, or the content within this book, without the consent of the author or publisher.

Disclaimer Notice:

Please note the information contained within this document is for educational and entertainment purposes only. All effort has been executed to present accurate, up to date, reliable, complete information. No warranties of any kind are declared or implied. Readers acknowledge that the author is not engaged in the rendering of legal, financial, medical or professional advice. The content within this book has been derived from various sources. Please consult a licensed professional before attempting any techniques outlined in this book.

By reading this document, the reader agrees that under no circumstances is the author responsible for any losses, direct or indirect, that are incurred as a result of the use of the information contained within this document, including, but not limited to, errors, omissions, or inaccuracies.

CONTENTS

Introduction	5
1. 3D printing: what you need to know	11
2. 3D Printer Models You Should Look Into	32
3. Top Must-Have 3D Printer Accessories	52
4. Picking Printing Materials	70
5. 3D Printing Software	108
6. First Print: Easy Step-by-Step Instructions	128
7. 10 Common 3D Printing Mistakes	142
Conclusion	157
References	163

A SPECIAL GIFT TO MY READERS

Included with your purchase of this book is the *3D Printing: 10 Beliefs to Stop Now* list. This printable will reveal the most common and surprising myths about 3D printing.

Click the link below and let me know which email address to deliver it to.

nataliebooks.com

INTRODUCTION

All right, so you've been sold on the idea of 3D printing.

Maybe you've seen friends, family, and strangers on the Internet who've bought home models of 3D printers and produced incredible things: toys and action figures, home decor and lamps, gadgets and tools, and even full-sized models of R2-D2.

Maybe you've seen 3D printers used in industrial and engineering settings, allowing for rapid prototyping of parts and hardware.

Or maybe you've seen the amazing plethora of advanced uses for this technology: everything from medical prosthetics to car parts to whole houses.

And now you're thinking, "I'd love to be able to create anything, right from the comfort of my own home."

Whatever got you hooked, you're hooked. You've been reading reviews of different models of 3D printers online; you've been deciding what you'll make with your new gadget. Maybe you've even already made a purchase.

And then it occurs to you: "I have no idea how to use a 3D printer."

Now you're online, looking at comparisons of filament types and wondering what the difference between PET and PETT is. Websites are talking about downloading STL files, but you're not even sure what an STL file is or how to open it. You don't know the first thing about 3D models or manufacturing plastic iPhone cases. You're in way over your head. And suddenly, you're not so sure that getting a 3D printer is a good idea.

Does any of this sound familiar to you? Are you feeling overwhelmed wondering if maybe 3D printing is just too advanced for you? That's what this book is for.

I'm Natalia Guzman. I've been studying 3D printing for seven years, and I think these printers are amazing machines! They have incredible potential for giving regular people like you and me the ability to create objects that are fun, useful, beautiful, or all three; I never get tired of seeing something appear where before there was nothing.

But my time using 3D printers has also taught me that it's not always easy to get them to work the way you expect. Unfortunately, it's very easy for beginners to start out enthusiastic about their new printers but then get stuck on some problem or issue, become frustrated, and give up altogether. I certainly went through periods like that, and I wished back then that I had a book of useful tips and tricks to get started and avoid basic problems. So now that I have seven years of experience under my belt, I decided to write the sort of book I used to wish I had!

I'm going to introduce you to the world of 3D printing, starting from the absolute beginning: you don't need to know how to do much except turn on your computer and look things up on the Internet. (If you already have some knowledge about 3D printing and are looking for more intermediate knowledge, you might want to take a look at the later books in our 3D printing series.)

We'll start with an overview of 3D printing: what it is, what it's for, and when it comes in handy. Then we'll discuss what you'll need to buy to get started, how to get started, and how to troubleshoot when things go wrong. Throughout the book, I'll give you step-by-step instructions and loads of useful tips and tricks.

I hope you'll find this book a useful reference, both to read as you're getting started and to refer back to in the future. 3D printing is a lot of fun; it's something you can tinker with endlessly as you find new things to print and figure out the best processes. But it's also something that can cause frustration when things don't work the way you expected, especially as you're just getting started. I hope to help you avoid the pitfalls that new owners of 3D printers face.

By the end of this book, you'll be printing with ease! You'll know how to find a file for the project you want to create and then guide that project through to completion. You'll easily create objects for fun, for use around your house, and more. And you'll have the knowledge necessary to avoid some common problems.

Are you ready to get started with 3D printing? Then let's go!

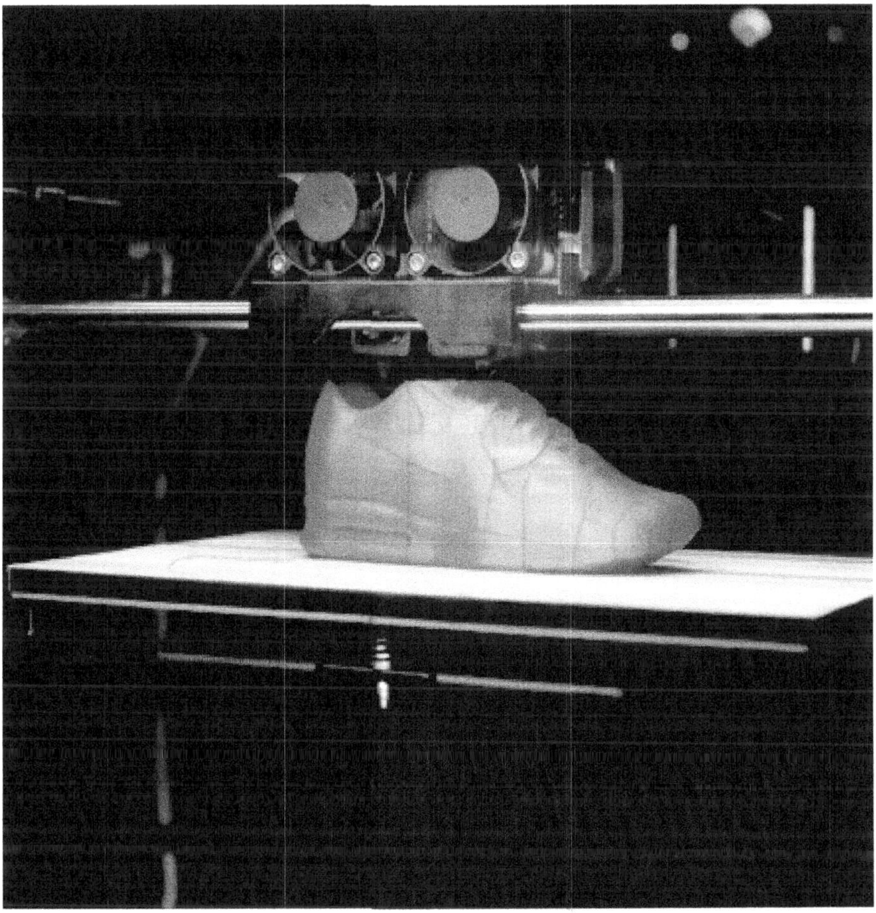

1

3D PRINTING: WHAT YOU NEED TO KNOW

Let's start, as the song says, at the very beginning.

WHAT IS 3D PRINTING?

Although the word "printing" might make you think of a computer printer, 3D printing only has a little bit in common with that.

Definition: *3D printing* refers to many different sorts of processes that involve a computer-controlled machine using some kind of material to form a 3D object. Often the material is set down in layers, and the layers often fuse to form a solid object.

In this book, we'll be talking about consumer 3D printers: generally small enough to sit on your kitchen table, affordable, and used by hobbyists.

But they aren't the only sorts of 3D printers; these machines have been used for many years for a variety of different uses.

It seems kind of fitting that the very first reference to an idea resembling 3D printing came from a science fiction story, given what a fantastic technology it is. Raymond F. Jones, in a short story called "Tools of the Trade," referenced a technology much like the 3D printing of today. But it wasn't until the 1970s that people started thinking much about the possibility of using this kind of process in the real world; a 1971 patent describes a rapid prototyping process using metal, and in 1974, British chemist David E. H. Jones suggested the possibility of 3D printing in a magazine column.

A Japanese researcher invented an additive manufacturing process for creating three-dimensional objects from plastic in 1980, but it never really caught on. Other groups and individuals around the world began working on technologies of their own throughout the decade. In the late 1980s, two important technologies were developed: Chuck Hull invented stereolithography, which involved a light-activated resin, while S. Scott Crump developed fused deposition modeling, which we'll talk about more later.

This is when the idea of 3D printing really kicked off. But back then, these machines were mostly used for rapid prototyping in industrial, manufacturing, and engineering settings.

What's rapid prototyping, you ask? Well, suppose you're designing a new case for a smoke detector you're manufacturing. You do all the design in a CAD (computer-aided design) program, and it all seems to look the way you want, but now what? You need to get a physical product in your hands so that you can test it and critique it and refine it. But how are you going to do that? Many engineering facilities don't have the necessary machinery to fabricate those cases. If you're at one of those facilities, you'll have to send that design off to another company that does have that capability. They'll fabricate your prototype and send it back to you. At that point, you'll test it and run it by other people, and based on your findings, you'll tweak the design, but that means you're going to need to send it out for fabrication again. Then they'll have to send it back to you . . . as you can see, this could involve lots of back and forth, lots of headaches, and lots of wasted time. If you have a 3D printer, though, you can quickly create that prototype in-house in a matter of hours and save yourself a lot of time and money. That's what we mean by "rapid prototyping."

This technology has a manufacturing use outside rapid prototyping, too; it can be used to create not just prototypes but also final products. It can be a useful manufacturing tool, especially when you don't need enough of the final product to make mass production worth it. Many early 3D printing technologies involved creating objects of metal, usually using powdered metal and lasers and the like. In industrial applications, such processes are often called "additive manufacturing."

Definition: *Additive manufacturing* refers to technologies that allow you to create 3D objects by adding material one layer at a time (this is in contrast to many traditional manufacturing methods, like machining or milling, where the material is removed). "Additive manufacturing" and "3D printing" are often used interchangeably, though the former is generally used in manufacturing settings, not for the kind of at-home hobbyist printing we're talking about here.

These days, 3D printing is used for all sorts of different applications, and they're not just using metal or plastic as their materials. The medical field has been using 3D printing for a while; as early as 2006, the Wake Forest Institute for Regenerative medicine 3D printed a scaffold on which they grew a replacement bladder for a patient. 3D printing also allows doctors to create implants and devices that are customized to specific patients; for example, in 2012, doctors in the Netherlands were able to create a titanium jaw to implant into a patient with a chronic bone infection. Another early example was in 2014 when doctors in Wales 3D printed plates and implants to help them reconstruct a patient's face after a motorcycle accident. 3D printers have also been used to create prosthetic limbs. Perhaps most amazing is the field of bioprinting, which focuses on the possibility of 3D printing organs and tissues, with a lot of really promising results so far. The possibilities of this technology for health applications are endless and will only get more complex and impressive as time goes by.

If you're too squeamish for the idea of printing tissues, how about eyewear? Some companies have started experimenting with 3D printing glasses frames. This means you can completely customize your glasses to fit your face since each pair is printed specially.

3D printing has also been used to create parts for cars and planes. Metal 3D printing allows for quick production of complex parts; these parts can often be designed to be more lightweight than their traditional counterparts as well. Toyota is experimenting with 3D-printed interiors, and many start-ups and research groups are working on or have created cars made largely or entirely from 3D-printed parts (well, except for a few parts like tires). I'm particularly excited about a car produced by Nanyang Technological University in Singapore that's partially solar-powered. And an American company called Local Motors has produced Olli, a 3D-printed, fully autonomous shuttle bus that's already been deployed at college and business campuses and on public streets in the USA, Italy, and Saudi Arabia. At present, these vehicles are generally not available for public purchase, although we hope they will be soon. In the meantime, it's predicted that the 3D-printed car market will grow to a $5.3 billion industry by 2023.

Museums have found 3D printing useful in creating replicas of objects in their collections; these replicas are used for study, for creating customized packing materials when objects have to be moved, and more. One interesting application is to allow visually impaired museum patrons to interact with a replica of these pieces by touching them—something that they wouldn't be allowed to do with a thousand-year-old statue. Previously, creating replicas of objects involved creating a mold, which could damage delicate surfaces; but scanning and then 3D printing these objects allows museums to keep them safe.

3D printing has even made it into the arena of food: 3D printers have been used to create vegan meat, and NASA is looking into 3D printing food for astronauts. And of course, the possibilities for creative food presentations are endless.

My personal favorite unusual application for 3D printing? Buildings. Large, specialized 3D printers can lay down layers of concrete, one at a time, to create a home for $10,000 or less in a matter of mere days. This method has created homes all over the world, including a whole community of houses for low-income families in Mexico, as well as hotels, schools, and offices. While this is still an unusual construction method, some companies are doing their best to move it forward, and some of these homes across the world are now available for purchase, in case you fancy a house of the future.

When most of us hear the term "3D printer," however, we usually think of consumer 3D printers, meant for use at home by hobbyists; these are relatively cheap and generally small enough to put on your kitchen table. These are the printers we're going to talk about in this book, and they are a relatively recent development. The idea of 3D printing really crept into the public consciousness toward the end of the first decade of the 2000s. In 2010, a company called Makerbot became the first company to exhibit a 3D printer (the Cupcake CNC) at the Consumer Electronics Show. More consumer 3D printers began to be available after that, many of them first funded through Kickstarter.

From there, it's gotten increasingly popular as a hobby. There are now numerous consumer 3D printers available at a variety of price points. The online community surrounding 3D printers grows every day. So there's never been a better time to get into 3D printing!

WHAT PRINTING TECHNOLOGIES ARE THERE?

As I said, there are loads of different print technologies: different machines, different processes, and different materials. In this book, we're going to focus on one called FDM.

FDM

For the kind of consumer 3D printers we're talking about, the most common technology is fused deposition modeling, or FDM. Though other options are available, FDM's ease of use and accuracy have made it a popular choice for consumer 3D printers.

(You'll also sometimes hear the term "fused filament fabrication," or FFF. A company called Stratasys has a trademark on the term "fused deposition modeling." However, their patent on the technology expired in 2009, which is why Makerbot was able to create its first consumer 3D printer that year. That printer, as you might have guessed, employed FDM technology.)

Definition: *FDM* is a 3D printing technology involving a thermoplastic material being extruded through a nozzle on a printhead. The printhead is controlled by a computer and deposits the material one layer at a time on the print bed to build the final product.

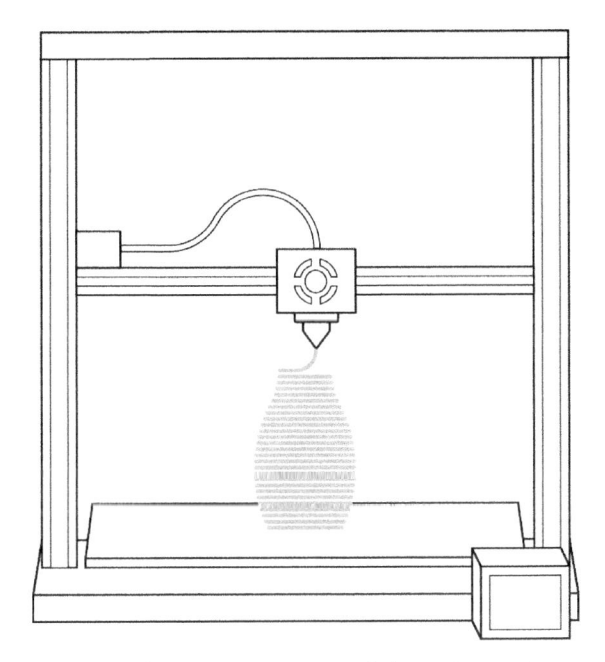

Fused Deposition Modeling (FDM)

Definition: *Thermoplastic* refers to a plastic polymer material that becomes pliable when you heat it but solidifies when it cools down.

Definition: The *print bed,* also known as the build surface, build platform, or the build plate, is the flat surface onto which the thermoplastic material is deposited.

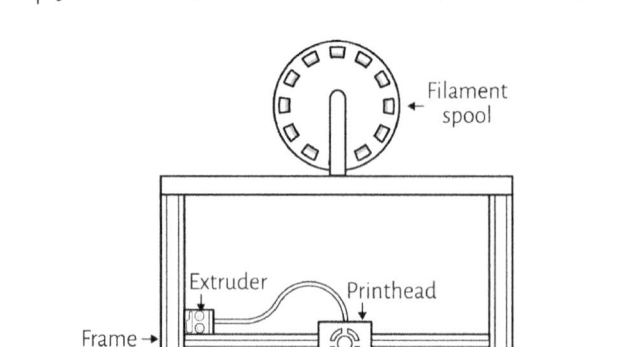

Parts of a 3D printer

So, to break it down into easy-to-digest steps:

- You start with a thermoplastic material in the form of a long continuous filament, like a string, usually stored on a spool.
- The filament is fed through an extruder and to the printhead. On the printhead is a nozzle, where the filament is heated up until it becomes soft and then extruded out.
- The printhead moves according to instructions you've given it via a computer file (using a language called G-code), depositing a single layer of the material on the print bed.

- Once a layer is done, the printer starts laying down a second layer. After being deposited, the hot material cools, fusing with the layer beneath it: hence the name "fused deposition modeling."

How long this takes will vary based on the size and complexity of the object; it can take minutes, hours, or, for particular complicated and high-quality objects, days.

Other printing technologies

There are loads of other options out there, often involving special materials and special lasers (and very expensive printers). For consumer 3D printers, after FDM, the most popular technology is probably resin.

Definition: *Resin printing* is a 3D printing technology where an LCD screen is used to harden a special resin in layers on a print bed.

Resin printing is a completely different process from FDM. It involves a vat of a special resin with an LCD screen underneath. A print bed is lowered into the vat, and the LCD screen projects the desired shape of the first layer of the object. This causes the resin to harden onto the print bed in that shape. Then, as with FDM printing, this process is repeated over and over to build the object in a series of layers.

Because it's such a different process from FDM, we're not going to cover it in this book. I bring it up partly because it's good to know what's out there—some people move from FDM printing to resin printing—and partly because you may sometimes hear about 3D-printed objects needing time to cure under a special light. I'm here to assure you that this is only true of resin-printed objects, not FDM-printed objects.

WHY GET INTO 3D PRINTING?

I have an easy answer to this question: because it's a good time! 3D printing is a great hobby for anyone who likes new gadgets and technology, who likes to tinker, who likes to create things.

Some people get into 3D printing strictly for the fun of it, because they like figuring out how to make things work. Other people print useful objects they'll actually use or give away or display around the house; my brother-in-law once surprised me with cookie cutters in the shape of my face!

3D printing is massively popular among board game and tabletop gaming enthusiasts as well: you can use your printer to create miniatures or board game accessories. In fact, one maker whose work I love is Ristow Designs (**ristowdesigns.com**, or check out their Etsy page), who makes incredible accessories for Settlers of Catan, including piece cups, cardholders, and whole boards made of magnetic, modular hexes to make gameplay easier and prolong the life of the tiles. I just love that as an example of using 3D printing to make things run a little smoother.

Another maker, YoCo Art, takes 3D printing to another level by making affordable art pieces and home décor that can be customized to different styles and colors.

Cosplayers and Halloween enthusiasts often 3D print props and accessories for costumes; this allows a level of accuracy and durability that would be hard to reach if you're just trying to carve the item out of styrofoam.

My friend from YoCo Art, (https://www.etsy.com/shop/YoCoArt?), uses 3D printing to design home decor items, which include custom pieces, sculptures, wall decorations, planters, and other decorative items. I invite you to visit YoCo's Instagram:

Building your own droids is a popular activity for Star Wars fans, with many builds involving 3D-printed parts; imagine showing up at a May the Fourth party with a full-sized R2-D2 replica! (Getting it to move and make noise, however, is a vastly different skill set that I'd be absolutely no help with.)

Buying a 3D printer is also a great way to get a child or teenager excited about invention, design, and engineering. What could be a better way to interest a child in technology than allowing them to choose a design for a toy and then having them watch as it appears before their very eyes? And there are other benefits too; I love this quote from Joel Leonard:

> "Kids learn not only how to design and produce new items. They also learn how to care and maintain equipment. Those skill sets can be transferred to numerous lucrative occupations including reliability engineering where we gave tremendous need and growth potential."
>
> — JOEL LEONARD, THE MAKERS' MAKER. MAKESBORO USA

Speaking of useful skills, 3D printers can be a great educational tool. In one paper by Torrey Trust and Robert W. Maloy

("Why 3D Print? The 21st-Century Skills Students Develop While Engaging in 3D Printing Projects"), teachers reported that students who used 3D printers as part of classroom learning developed a number of useful skills, including creativity, technology literacy, problem-solving, self-directed learning, critical thinking, and perseverance. 3D printing in the classroom could open students' eyes to a whole new world of engineering, design, and technology.

But 3D printing doesn't have to be purely for education and entertainment! Many people have found ways to make money off of 3D printing. A quick search for "3D printed" on Etsy brings up hundreds of thousands of products for sale, including everything from custom dog food dishes to jewelry to busts of famous people. I have a friend who licensed the characters from a TV show, created action figures of them, and set up a website to sell them online. Some people also do custom prints for customers who don't have 3D printers of their own. Of course, you'd need a pretty decent printer for this, but if you have one, you could start advertising online and look for people who'd be willing to hire your printing services.

As you can see, 3D printing can be done for fun, for education, and for profit. And when you get involved, you can find yourself in a world of "makers": tinkerers like yourself who like exploring new ideas and technology, who like to push the limits of what's possible and explore new applications for the technology, who like to create. One of the things that I really love about 3D print enthusiasts and makers, in general, is the willingness to share; many 3D printing technologies and projects are open source and based on open hardware. In fact, the project that really got a lot of people excited about personal 3D printers, the RepRap project, was an open design project. We're all tinkerers, and we want to help you tinker too. So embrace the DIY spirit and join us!

SO 3D PRINTING IS EASY, RIGHT?

This is a good place to stop and give you a gentle, friendly word of warning. 3D printing is not, in fact, easy. There's an art and a science to it, and it'll take a bit of effort on your part to figure out how to best use this piece of complex machinery and create high-quality final products.

3D printing is a great hobby for people who like to tinker with stuff. If you like gadgets and new technology, if you like fiddling with things to figure out how to make them work, you're going to love this! But if you're getting into 3D printing purely to create succulent planters in the shape of an octopus to sell on Etsy, a word of caution: you're not going to buy a 3D printer, set it up, and immediately start producing perfect products. It'd be great if it worked this way, but it just doesn't. If you're not sure you like the idea of messing with a new gadget and figuring it out—if you're super focused on getting a great final product as soon as possible—you might find the first little while of your printer ownership very frustrating. And maybe, just maybe, this hobby just isn't for you.

But if you do like the idea of tinkering and fiddling, you're going to love 3D printing! I'm excited for you to start down this path and learn how much fun it can be to own a 3D printer.

Here are a few tips for getting started.

Have realistic expectations

Even if you make mistakes, it's okay! Don't be discouraged if your first few prints don't work out. In fact, much like the proverbial first pancake, your first prints probably *won't* work out. Expect to waste a bit of filament and to spend the first few days just messing around. Enjoy the process! Enjoy the things you're learning! And don't give up when things don't go well at first.

Get some support

There is a large online community of makers and people who are very excited about 3D printing. You might consider joining a group on Facebook or elsewhere online. You'll get a lot of encouragement, a lot of good ideas, and most of all, you'll have people you can turn to when you've got a problem you can't figure out. Everyone in these groups was once where you are now; they had the same issues starting out. You can even find online groups where people use the exact printer you have, making them extra useful as a support system.

Also, check out YouTube; you'll find more videos giving advice about 3D printers than you can shake a 3D-printed stick at. There's something about seeing something in action that really helps you understand it, isn't there?

Be excited

This is a hobby for enthusiasts; therefore, you need to be an enthusiast to really get a lot out of this hobby. It will take time, and it will take effort and problem-solving and troubleshooting, and occasionally you'll want to tear your hair out when a print fails three hours in, and you have no idea why.

But it's worth it! It is so, so worth it, as a way to have fun and exercise your creativity and increase your tech-savviness and create fun or beautiful or silly or useful things. You will love this hobby when you get into it. And though it can seem a little mystifying at first, you will figure it out! Read this book, do your research, watch some videos, level your print bed, and you will do just fine.

GET IT RIGHT IN 3 STEPS.

Want to make sure you get your 3D printing off on the right foot? I have three steps I like to recommend to beginners to increase their chances of success.

- Choose the right printer
- Choose the right materials

- Choose the right accessories

That's what we're going to talk about in the following chapters. After that, we'll go through your first print job. Finally, I'll give you some common mistakes and how to avoid them.

So, feeling prepared and excited? Then let's get started.

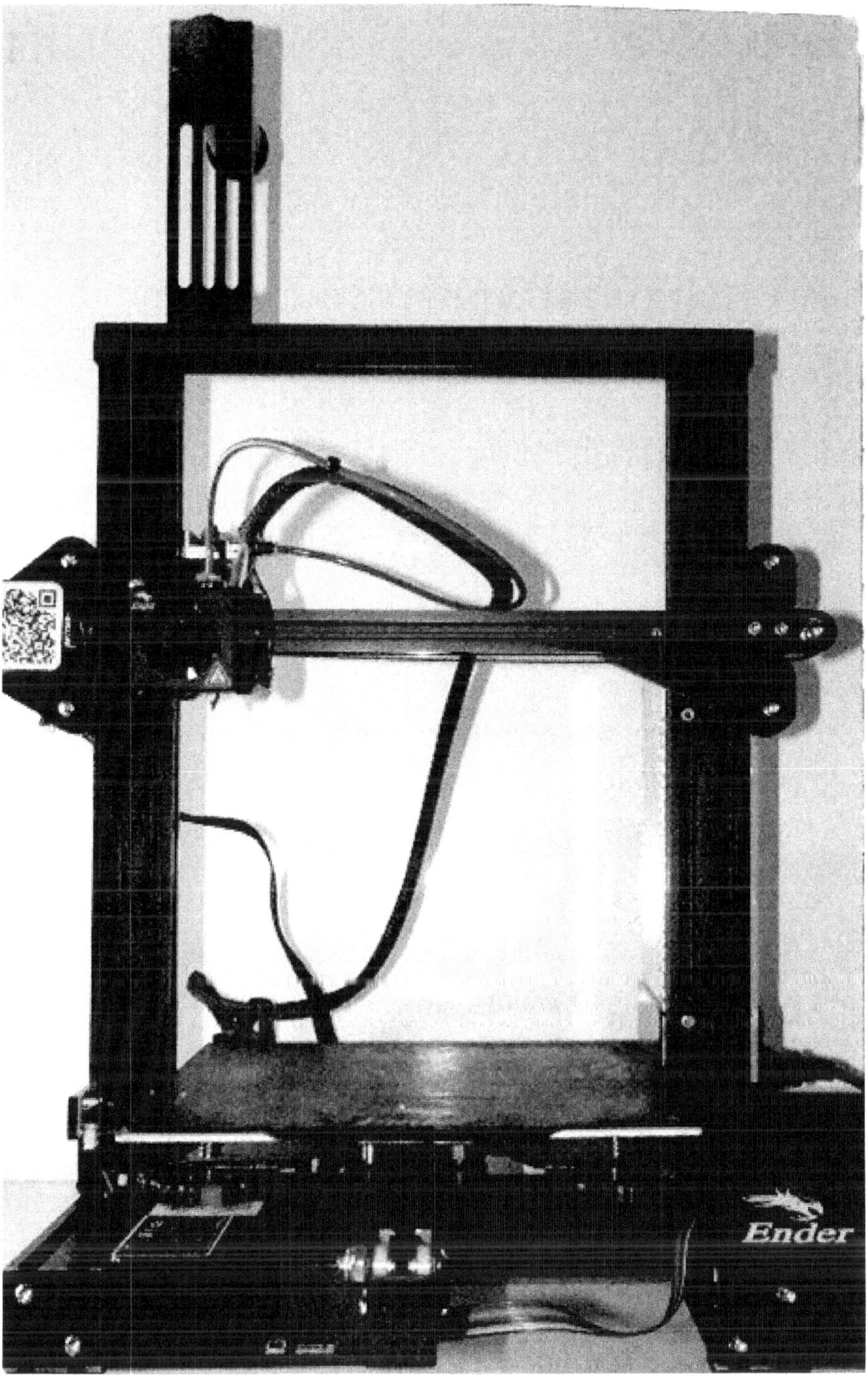
Ender

2

3D PRINTER MODELS YOU SHOULD LOOK INTO

Obviously, the first thing to do before you can 3D print anything is to choose a 3D printer. Maybe you've looked already and been bewildered by the wide variety of printers available—and the wide variety of prices you could pay for them. Luckily, you've got someone to guide you in picking a printer (hint: it's me).

But before you can make that decision, you need to think about your purpose in 3D printing. Because there's a lot of variation in what consumer 3D printers are out there, right? They are going to vary in price, ease of use, customizability, quality, and a load of other factors, and as with many things, there are going to be trade-offs. If you want the cheapest possible printer, the objects it produces will likely not be as high-quality as what you'd get from a much nicer printer.

3D PRINTER MODELS YOU SHOULD LOOK INTO | 33

So you need to decide what it is you really want this printer for, so you'll know which qualities to prioritize: if this something you just want to do for fun or to amuse and educate your children or students, you can probably choose a cheaper but lower-quality printer and be fine. On the other hand, if you're planning to create scale models of the Arc de Triomphe to sell on Etsy, you're probably going to want a printer that'll produce better results to keep your customers happy.

So take a minute and get serious with yourself as you decide what you want to use your printer for, and then get even more serious with yourself as you take a look at your budget.

And finally, keep in mind that the choice you make now doesn't have to be the choice you make forever. As is often the case, when you buy something you end up using a lot, you may use it for a while and come to realize that you value certain features or abilities more than you thought you would. When that happens, you may be able to update or upgrade the printer you have (figuring out ways a printer can be upgraded is one advantage of open-source designs being constantly tinkered with by a community of makers), but you also may buy a new one at that point. You usually just won't know what's most important to you at first, which is why it's generally best to start with a reliable, affordable printer with a large community backing it. You can start there, get your feet wet, and then try something new in the future if you want.

Got all that? Then let's look in-depth at some of the considerations you need to take into account when choosing a 3D printer.

THINGS TO LOOK FOR

Build quality

Not all printers are going to produce the same quality of products. Cheaper printers are often going to give you a lower level of detail in the final product. They will also likely be less reliable and durable long-term. So if you wanted to produce something with really fine and precise details and want the printer to last you for ages, a cheaper printer might let you down.

Material compatibility

We're going to talk more about the different print materials you can choose from later. For now, just be aware that different materials are going to have different needs as far as heating goes. Generally, printers will be able to work with really common materials, like PLA or ABS. But not every printer is going to work with more unique materials that might require really high heat. And as with above, generally, the cheaper models are going to have fewer capabilities there.

Support

I'm talking here about two kinds of support: customer and community. Customer support is essential, of course; you want to buy from a reputable company that will help you if something goes wrong. But it's equally important to buy a printer with a lot of community support. The more popular a printer is, the more people will be talking about it online, and the more likely it is that you'll be able to find a community online of people willing to help you when you have questions. So if you're considering an obscure or uncommon model, keep in mind that you'll be sacrificing some of that community. At least here at the beginning, when you're just dipping your toes into 3D printing, you might be best served with a popular printer with loads of people online talking about it.

TYPES OF FDM PRINTERS

Just to make this as confusing as possible, there are several different types of FDM printers. The differences between them tend to revolve around the way the printhead moves to create the final product; as I mentioned above, sometimes the print bed moves as well to help the process along.

Several of these types of printers fall under the label of Cartesian printers. Now, if you're good with math, or you just remember your high school math classes, your ears might have pricked up at the term "Cartesian." Yes, this name is a reference to Cartesian coordinates, a system in which a point's location can be described by numerical coordinates that give its distance from reference lines: the x, y, and z axes.

Chances are, if you can picture a 3D printer in your head, you're picturing a Cartesian printer. These tend to have that familiar boxy shape, with frames with lots of right angles so that the printhead can easily move along the x, y, and z axes.

Definition. *Cartesian printer:* A type of 3D printer where the printhead is directed using Cartesian coordinates.

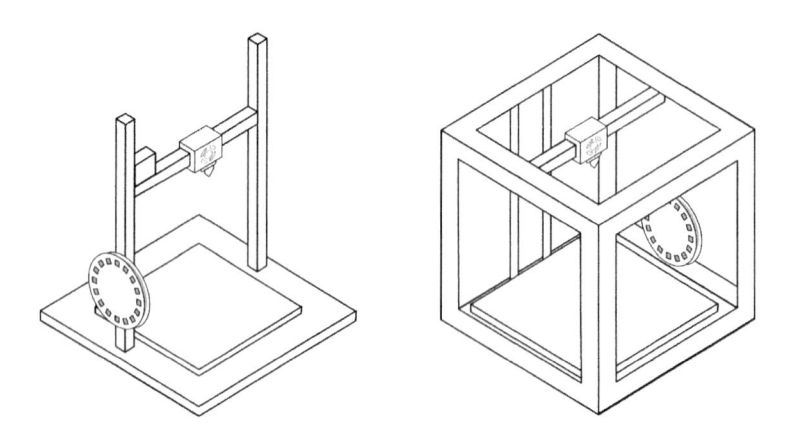

Common Cartesian printer styles: open (left) and enclosed (right)

These come in a few different configurations. Generally, the print head is on a gantry and can move along two axes, while the third axis of movement comes from the print bed moving. It may be open, as in the image on the left, or fully or partially enclosed, as shown on the right.

Cartesian printers are the most popular type of 3D printer, which makes them a great choice for beginners because with so many people owning them, you're going to find loads of support and information online. They are also often fairly affordable and easy to use, so the printers I recommend in this chapter are all going to be Cartesian.

Some types of these Cartesian printers, like H-bot and CoreXY printers, use belts to move the printhead; with these printers, often the print bed itself also moves up and down to include all three axes of movement. CoreXY, in particular, is making a splash in the 3D printing community, as they tend to be quite accurate and stable; the unique belt system leads to fewer vibrations.

There are a number of other unique but less common printer types, like SCARA, which uses a printhead on a robotic arm, or delta, a sci-fi–looking device where the printhead is suspended from three arms that can be moved up and down in different configurations to move the printhead anywhere it needs to go. One particularly unique specimen is the belt printers, where the print bed is a conveyor belt. This means that if you kept the conveyor belt moving and didn't run out of filament, you could theoretically 3D print something that's miles long.

There's one last type of printer that's completely different from what we've just talked about; these are polar printers, and instead of Cartesian coordinates, they use polar coordinates. They also have the printhead on a single arm and employ a spinning print bed. These aren't very common, but they're definitely interesting to see in action!

Some assembly required

There's one last thing we need to talk about before we can get onto individual printers: how much assembly do you want to do?

As I've mentioned before, the world of 3D printing is all about that DIY, get-your-hands-dirty sort of attitude. So some people really want to tinker with the printer; they want to be elbow deep in the metaphorical or literal gears of their machines. If that catches your fancy, you can find printers that are fully open source, open hardware, and customizable; these printers are designed for you to fiddle with! They also tend to use open-source software.

(Many such printers are the descendants of that original RepRap project I mentioned previously. Fun fact: "RepRap" stands for "replicating rapid prototyper," because the original idea, started by Dr. Adrian Bowyer at the University of Bath, was to create a low-cost 3D printer that could produce some of its own parts, thereby replicating itself. You can use a 3D printer to 3D print a 3D printer. Just think about that and marvel . . . or panic about the inevitable rise of the machines. Or both!)

That kind of thing isn't for everyone, however, so you may be pleased to know that there are also printers that are ready out of the box, requiring little or no assembly. They're a little less open, a little less customizable, but they're also often a little more user-friendly.

So take stock of what you want here: do you want to be able to customize your printer? Do you like being able to pull things open and see how they work? Or do you prefer things to just work right out of the box? Keep the answer to that question in mind as we go through the rest of this chapter.

Now that you have a good background in FDM printing technology and different factors that might affect your purchasing decision let's start talking about printers you might want to consider for your own use.

SOME PRINTERS TO CONSIDER

We're going to talk about a few different categories of printers here, ending by talking about my personal favorites.

Most affordable

First things first: always be aware of what you're getting into when you make rock-bottom prices your number one criterion for making a purchase. In many cases and many fields, buying cheap equipment becomes a self-fulfilling prophecy: you think, "I don't know how seriously I'm going to take 3D printing (or photography, or learning the violin)," so you buy the cheapest possible 3D printer (or camera, or violin). Because you have cheap equipment, you find the process of using it to be less than pleasant, and it produces mediocre results. You get discouraged about the fact that your 3D prints (or your photographs, or your sonatas) aren't very good, and you lose interest in your new hobby and give it up. And then you think to yourself, "It's a good thing I didn't waste money buying good equipment."

Obviously, sometimes going cheap is the only viable option. But I would like to recommend that where it's possible, you splash out a little more cash to get yourself the best 3D printer you can. Think of it as an investment in your future enjoyment. And wouldn't you rather pay a little more upfront than buy a cheap printer and then three months later have to buy another printer when you realize the cheap one can't do what you want it to? Especially when the difference between a good printer and a bad printer can be as little as $100 or less.

That being said, sometimes, for a variety of reasons, cheap printers are the best option. And luckily you've got some affordable options! One of the great things about 3D printers becoming so widespread is that there are now quite a few manufacturers of 3D printers, meaning there's more competition among manufacturers, driving prices down.

Now, that's more than enough of a preface: let's talk about affordable printers. One company you should consider is Anet, a Chinese company selling affordable printer kits. Anet does two product lines, the A line, and the ET line. Though it costs a little more than the ET line can, I'm going to recommend the **A8**

Plus, which many people agree is one of the better printers you can get at its price point (at present writing, this printer costs a little over $200 on their website, although you can find pretty cheap deals if you poke around a little bit online).

This printer has the common open layout I mentioned earlier, where the printhead is housed on an arch or frame over the print bed, and all sides of the print bed are open. This means you have a pretty large build volume (220 x 220 x 240 mm) compared to other cheap models (in particular, many printers in this price range are "mini" printers, so as you might guess from the name, they're pretty small). It's also got a heated print bed, which is common at higher price points but not guaranteed in such a low-cost printer, so that's pretty nice. (We'll talk more about bed adhesion later, but just know that a heated print bed is a great thing to have, especially if your print material is ABS.)

In terms of print quality, most people agree that this printer does all right straight out of the box, but you can get pretty decent results once you've started tinkering with settings. And one big plus for this A8 Plus is that the Anet models are reasonably popular, so you're going to find online support and videos and Facebook groups.

All right, there's what's good. Before we talk about what's bad, let's talk about what could be good *or* bad, depending on how you feel: the A8 Plus is a kit, and you will have to assemble it yourself completely. Users report this taking anywhere from 6 hours to several days, depending on their level of technical know-how. Now, you might quite like this if you're a technical sort of person; it's a great way to come to understand how 3D printers work so that in the future, you'll be able to upgrade and troubleshoot easily. On the other hand, this might be a little overwhelming for a beginner.

Finally, what's bad: well, it's a cheap printer. The frame is made of acrylic, not metal, which means that this is not going to be the most stable and sturdy printer, which can lead to bad prints because of less stability (most printers above this price point have metal frames). Some users print supports and extra parts to stabilize their A8 Plus printers.

The design has some flaws, including exposed electronics, and can have overheating problems, occasionally serious ones. Many users fix these problems by upgrading the firmware and hardware. There's also no automatic bed leveling, which is something we'll talk more about later. Finally—and this isn't a knock on the A8 Plus in particular, because many printers have this problem—the open layout plus the heated print bed could be a problem if you've got small ones at home who might like to grab at things that might burn them.

So, if you've got that DIY spirit and that can-do attitude, and you're willing to do the assembly and make the upgrades as needed, this can be a great little budget printer for you. If you don't want to do quite so much assembling and upgrading, though, you may want to keep looking.

Easiest to use

Maybe you read that last section and thought, "I don't want to do nearly that much assembly and upgrading; I want to open the box and get started." Well, in that case, first, I would tell you that being willing to do upgrades and assembly, being willing to get your hands dirty, is going to be a great choice in the long run. You're going to learn how 3D printers work, and you're going to be ready to make the upgrades needed to make your 3D printer even better.

But I get it: this is your first 3D printer, and you want to learn how to 3D print before you get bogged down in the esoteric details. Totally understandable! In that case, for a truly user-friendly experience, you should consider the **Flashforge Adventurer 3**. With this printer, the printhead and print bed are enclosed (that it's enclosed is an advantage if you're worried about kids burning themselves on a hot print bed).

The Flashforge Adventurer 3 is basically the opposite of the A8 Plus. It's designed to be as easy to use as possible: it arrives fully assembled and ready to go, so you'll be printing shortly after opening the box. It's tidy and fully enclosed, and it comes with a camera so you can watch the process from your computer.

The printhead is detachable, and the heated print bed is flexible, which means that it's easy to get your final product off it: you just flex the platform a little, and it'll pop right off! This printer also runs fairly quiet, and the print quality is quite good.

This all sounds pretty great, right? So what's the catch? Well, there are a couple. The build volume is small, to begin with: 150 x 150 x 150 mm, while the A8 Plus is 220 x 220 x 240 mm, and the Creality 10s, which we'll talk about later, boasts a whopping 300 x 300 x 400 mm. The price may give you pause as well; they'll cost you between $400 and $500.

And lastly, the very fact that this printer is such a tidy little self-contained package could be considered a con; it's pretty much a closed system, and you can't really mod or upgrade this the way you can a lot of other printers.

So: great if you want something easy to use and reliable; less great if you were hoping to be able to tinker with it. So if you fall into the first category, this could be a great option for you.

Best pro printer

And now for something completely different: maybe you want it all. Maybe you want the build volume, the quality, the reliability. Maybe you want something of professional quality, and you're willing to pay the big bucks to get it.

In that case, may I recommend the **Raise 3D Pro2**? This is a beast of a device with quite a lot of high-end features; it would, as the name suggests, be appropriate for professional applications. It features an enclosed, boxy design, but unlike the Flashforge Adventurer 3, this doesn't mean a smaller build volume: it boasts an impressive 305 x 305 x 300 mm capacity. As with any printer at this price point, the print bed is heated, and the printer can handle a wide variety of printing materials. If the power goes out mid-build, all is not lost; it'll resume the print job when the power comes back on! (And when complex print jobs can run for days, that's a vitally important feature.) It's easy to use, and it's fast.

A really interesting feature on it that we haven't seen yet in our tour of printers is the dual extrusion: the print head has two nozzles on it. "Now why," you may be asking, "would I want two nozzles?" There are two main reasons: one, you could use it to create two-color designs.

Two, you could use it to create designs out of two different materials. This is great if you need to print support structures: sometimes your prints will have overhangs and bridges, and to keep the overhanging part from sagging, you'll need to include support structures that prop the part up from beneath; you can then remove the support structures when the print is done. But removing them can be a pain, particularly if you're trying to make it so you can't tell there were ever any support structures attached. One solution: use one of the dual nozzles to print support structures with a material like HIPS, which can later be easily dissolved away in a substance called limonene. Pretty handy, right?

So what's the downside to the Pro2? That should be obvious: the price. You may not be ready to spend a few thousand on a 3D printer, and I don't blame you if you're not. But if you're willing to spend the money—and especially if you want to use your printer in any sort of professional industrial application—this is an excellent choice.

SOME OF MY FAVORITES

I've owned five 3D printers in my time. I think each of them has some great features, and each could be the right choice for certain people.

Artillery X1

Artillery is a relatively new company, but they've put out a respectable sub-$500 printer in the **Artillery X1**. It comes preassembled and does a decent job printing. It's reliable with a good-sized bed (300 x 300 x 400 mm), and it runs very quietly. The print bed does heat up unevenly, though, and some users have reported concerns about the long-term reliability of the electronics and cabling. However, so far, I'm a fan of this printer.

Anycubic Mega Pro

Another printer I own is the **Anycubic Mega Pro**. One of the cheaper printers we've talked about on this page—at the time of writing, less than $400—has a very similar form factor to the A8 Plus we discussed earlier; it even has a similar build volume, at 210 x 210 x 205 mm. It's got loads of useful features—like a sensor that notices when you're about to run out of filament—that are pretty great at this price point. It's an excellent little machine as it comes, and the growing community of fans online can offer you ideas for upgrades to make the experience even better.

What makes these printers so remarkable, though, is the fact that it doubles as a laser engraver. Just swap out the printhead for the engraving attachment, set the item you want to be engraved on the print bed, and you're good to go.

The big downside, if you're not in the mood to get your hands dirty, is that this printer does require you to assemble it.

Still, if you could see yourself wanting to laser engrave things, this is an excellent 2-in-1 tool.

Prusa i3

One of the better printers I own is the Prusa i3, and I am not alone singing Prusa's praises. This company has been hugely influential over the last decade; its open-source **Mendel** designs, particularly the **i3** version, are the ancestors of many of the consumer 3D printers we see today. (Look at a picture of the original i3 and then of nearly any of the printers we've talked about so far, and you'll see the family resemblance.) The company released the first **i3** printer back in 2012, close to its founding, and absolutely changed the game.

That original i3 design has been tweaked to near perfection since that first release, and it has the awards and accolades to prove it. The latest version is the i3 MK3S+ (quite a mouthful, I know), and it has a host of useful features, like a heated print bed, a low-filament sensor, the ability to pick up where it left off if the power goes out during a print, and a feature they call mesh bed leveling, where the printer will try to compensate for imperfections in the print bed. It also comes with removable print sheets in a variety of textures, which make removing the final product a snap.

It's consistent, it's reliable, the final products look great, and this is a workhorse: it just keeps going and going.

So what are the downsides? Well, it's not the cheapest, with pre-assembled printers running for around $1000. You can get that price down to $750 if you're willing to assemble it yourself. You'll also find that the build volume is only middling, at 250 x 210 x 210 mm.

Still, the Prusa i3 is a classic, and there's a reason. If you're willing to fork over a little extra cash, this is a fantastic printer.

Creality 10s Pro V2

Pricier than the budget models we've been discussing but cheaper than the Prusa i3, the **Creality 10s Pro V2** is a sturdy, reliable workhorse. I already mentioned that large build volume—300 x 300 x 400 mm—due to its open design. It has the features you'd expect at this price point—heated print bed, sturdy metal construction, the ability to use a variety of materials, low-filament sensor, the ability to resume a print job after power has been lost—and a few fun new ones, like an auto-leveling feature (and if you've ever had to manually level a print bed, you'll know that's not nothing). It runs quiet, and the final products look great.

The Creality 10s Pro V2 also comes mostly assembled; you just have to attach the gantry, plug in a few wires, and you're good to go!

So, what are the downsides? Well, it's not the cheapest, and other users have complained about the lack of good documentation, though their customer support is excellent. I've also seen users complain that it can be a little difficult to get the final product up off the print bed.

But really, that's worth putting up with for a printer this good. It just works.

Ender 3 Pro

My last 3D printer is also manufactured by Creality; it's the **Ender 3 Pro**, and it's definitely near the top of my list. And it's not just me; this is a popular printer and is widely recommended by people as one of the best printers for beginners. It's just a perfect balance of price and performance, at around $300; if you don't have a lot of money to spend, this printer is going to give you a lot of bang for your buck.

It's got the usual list of features at this price range—sturdy construction, heated print bed—and a decent build volume at 220 x 220 x 500 mm, due to that familiar Prusa i3 layout. The print bed has a magnetic layer on top that can be removed; it's bendable, making it easy to pop finished products right off it. It can use a respectable selection of materials, and the final products are high quality! And it's customizable, which makes it very popular in the maker community.

Of course, at this price point, you're sacrificing some swanky extra features, like the resumption of printing after a power outage and auto-leveling. And, like a lot of Creality printers, it comes only partly assembled; you'll have to connect a few parts and plug in a few cables.

But, for this price, you couldn't do better for yourself than the Ender 3 Pro.

So what do I recommend?

I highly recommend either the Ender 3 Pro or the Creality 10 S Pro V2. I think the Ender 3 Pro prints better quality products, but really it all comes down to your needs: do you need a higher build volume? Go with the Creality 10. Are you on a budget? Go with the Ender 3 Pro.

Really, I couldn't tell you which one I like best. If I could only keep one of the two . . . I'd be thrilled with either one. They're both excellent choices.

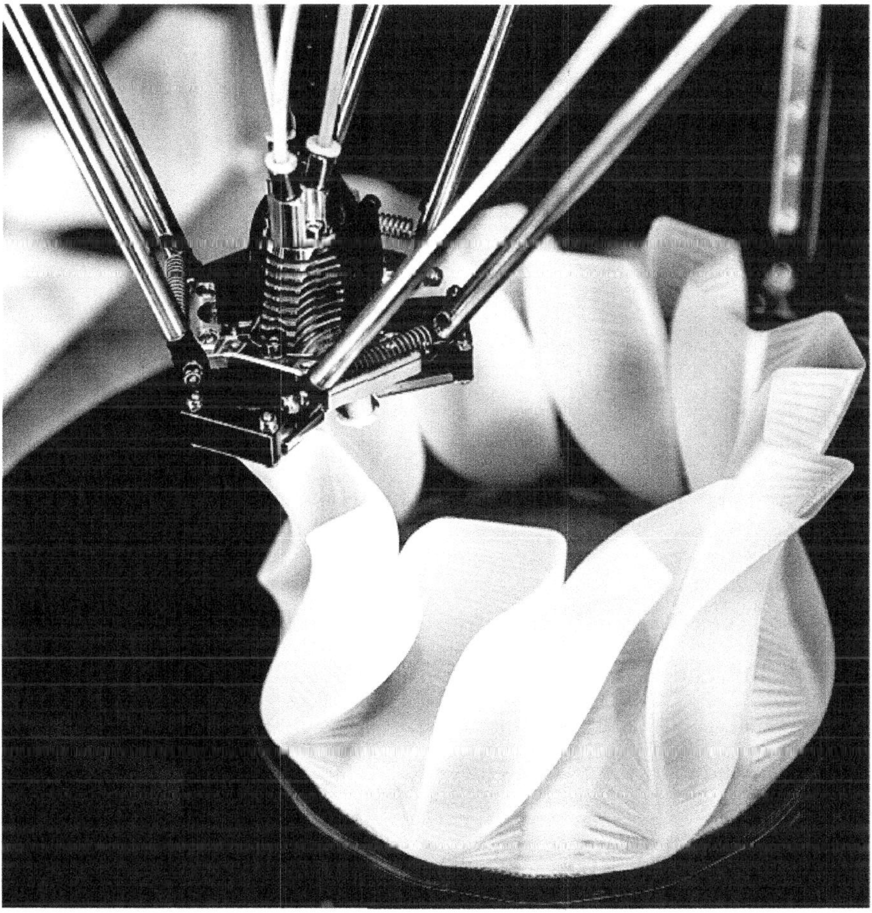

3

TOP MUST-HAVE 3D PRINTER ACCESSORIES

So you'll have noticed that in the last chapter, I talked a lot about upgrading your printer. What, you may be asking, does that mean? And why would you need to upgrade your printer?

If you've chosen well, your printer should work decently well as is. But, since it's a hobby printer—especially if you chose a budget hobby printer—well, it's just not going to be as fancy as a professional printer would be.

But you don't have to put up with that! This is 3D printing, where DIY tinkering is not only allowed; it's encouraged. If you're willing to spend a little money, get your hands a little dirty, or both, you can upgrade your printer to produce better results and a more pleasant printing experience.

(Keep in mind, though, that different printers are going to allow different amounts of customization. If you get one of the open-source, build-it-yourself kits, you can basically access every piece of the printer and do what you want with it. With a more user-friendly, fully-assembled printer, like the Flashforge Adventurer 3, it's already sort of a self-contained unit, and upgrading it isn't nearly so easy. That's the point of that printer, after all: that they've done all the work and the thinking for you. So how much upgrading you can do will definitely vary based on what you buy.)

The upgrades you can do fall basically into two camps: stuff you print yourself and stuff you buy from somebody else.

YOU MEAN I CAN PRINT MY OWN UPGRADES?

This is one of the coolest things about getting into 3D printing. In the spirit of those RepRap self-replicating printers, you can actually print parts that you can snap onto your own printer!

You may remember me mentioning this possibility when we were talking about the Anet A8 printer; with some of the cheaper printers with acrylic or other less-than-sturdy frames, the frame might wobble during a print and negatively affect your final product. Luckily, the solution is right there in front of you: you can print a brace, a supporting frame, or other stabilizing features for your printer. And you don't have to buy them! You can use your 3D printer to 3D print parts to improve your 3D printer!

It doesn't stop there; you can print all sorts of things that would be helpful for your printer! You can print pieces that make it easier to change your filament spools, or that guide the filament nicely into the extruder, or that you can run the filament through to clean any dust off of them. You can print new mounts for your printhead that can give higher accuracy or speed in your prints. You can print pieces that hold belts in place or keep up the desired amount of tension on them. You can print elements to help you organize cables and wires or cover up exposed electronics (very helpful on some of these cheaper printers). You can print vent covers to keep dust from falling into your vents. You can print gadgets to help your printer run more quietly. Really, the possibilities are nearly endless.

I'd love to tell you what upgrades you ought to consider for your printer, but I can't. Every 3D printer is different; each one is going to have different strengths and weaknesses and, therefore, different needs. And your print needs may be different from someone else's! Maybe you are printing things that need really precise, sharp corners, and the printhead mount that came with the printer just isn't going to cut it, but someone else who prints only flowerpots with the same printer has no problems whatsoever.

So instead of telling you exactly what you need, I'll give you a piece of advice: find an online community of people who use your precise model of 3D printer. Find videos or forums. Search Thingiverse for "[your 3D printer model] upgrades," and look at the models that have lots of makes and feedback, and figure out whether it successfully solves a problem you find yourself having. Or just wait until you've been printing for a while; you'll soon come to realize what weak spots your printer has, and then you can search for a solution.

And then you can print that solution and marvel at how nice it is that you can make your 3D printer work better just for the price of filament.

(Remember what I said earlier, though, about certain printers not really having as many options for upgrades. You may search and find that there's not much you can do to change your printer. In that case, I hope your printer works great as it is.)

WHAT ABOUT ACCESSORIES THAT CAN'T BE PRINTED?

Not every problem can be fixed with 3D-printed parts. However, as amazing as these machines are, you might find yourself needing a solution that they just can't produce. In that case, there's a wide variety of great accessories you can purchase online and use with your printer.

When you're looking at these parts, keep in mind that for many of them, they are going to be different for each printer. Each printer has different dimensions and a different configuration, so it's going to require a different accessory; these are not one-size-fits-all parts you're buying. So instead of telling you exact product names and part numbers, I'm going to go over some of the popular accessories and why you might want them. If you think one sounds good to you, you can go find which version of that product will work best with your printer.

Note that there are way more accessories and upgrades available out there than what I'm going to talk about. You can seriously go all out on this stuff, with people upgrading motherboards, motors, and more. That's outside the scope of this book, though, as this is aimed at beginners; my plan is to focus on things that are easy to integrate and will give you big results. So just keep in mind that there's more out there than what we're going to discuss here.

Make sense? Then let's talk about some useful accessories for printers.

Print bed

You'll see a lot of accessories out there aimed at the print bed, and there's a reason for this: your print bed can have a huge impact on your final product. One big reason for this is bed adhesion.

Definition: *Bed adhesion* refers to the extent to which the printed material adheres to the print bed. Too little adhesion can mess up the lower layers, as the outer edges may cool faster than the inner layers and start to peel up from the bed, warping the layer. Too much adhesion, on the other hand, can make the final print difficult to remove.

In later chapters, we'll talk about DIY fixes for getting bed adhesion just right, but you can also purchase a couple of accessories to help with this problem.

One option to consider is adhesives. Often in the form of glue sticks, these are products that you can spread on the print bed that will help your first layers stick properly but will then easily release the print when it's done. For instance, products like Magigoo are sticky when put onto a heated print bed, but they lose their stickiness once the platform cools down. At that point, it's easy to remove the finished product.

You can also get new surfaces for your print bed. Many people prefer to put glass surfaces on their print bed, finding them to be more consistently flat than many build surfaces, which often also makes it easier to remove the final product. They're also easier to clean than a lot of other surfaces.

You can also buy special metal build surfaces that sit atop your print bed and, thanks to their texture or chemical makeup, provide excellent build adhesion. You can even buy flexible build surfaces; these make removing prints a snap because when the printing is done, you simply flex the build surface, and the print pops right off!

(Certain printers come standard with these interesting and useful print bed options, so that's something you should keep an eye on as you're considering printer models.)

Enclosure

So a 3D printer works because we use thermoplastics, which become more liquid when they get heated up, allowing us to push them through an extruder and layout the desired final shape; when they get cool, they firm back up. We've talked about this already, right? The bit that we haven't talked about is that not every thermoplastic reacts in the same way when you cool it down. Some have no problems. Others definitely have problems, specifically when the different layers and different parts of the print are at different points in the cooling down process. This can cause the lower layers, in particular, to have problems; it's common to see prints curling up a bit at the corners or other warping.

So what do we do about it? Well, one thing that can help is to control the printing environment. An enclosure that covers up your entire 3D printer can keep the air inside a consistently warm temperature while the printing is happening, leading to less warping because you can control the process.

(Two other perks to using an enclosure: it keeps dust off your print while it's being printed, and if you've got little kids around, it can help keep little hands from grabbing hot things they shouldn't.)

This is something that a lot of people choose to DIY; really, all you need is something bigger than your printer that can cover it up. People have jury-rigged boxes and plastic sheeting and—my personal favorite—IKEA end tables into very serviceable enclosures for their printers.

However, you might prefer an enclosure that someone else has made for you, both for the sake of ease and because these enclosures often offer features like fireproof materials (just in case something goes wrong with a print and you're not close by). If you look online, you'll see that many of these enclosures are designed to fit specific printers, so if you're going to get one, make sure you're getting the right one.

You'll also want to keep in mind that certain printing materials like ABS will give off odors, so whatever enclosure you end up using needs to allow you to ventilate properly.

Nozzle

One part that many users choose to change out for a variety of reasons is the nozzle on the printhead. The process of changing out a nozzle can usually be done just using a wrench, but it's good to read up in your user manual and online to make sure you know the best way to do it on your printer.

So why would you want to change out the nozzle? There are a few reasons.

- **To clean the nozzle.** Nozzles can get clogged and require periodic cleaning out. If you had a spare nozzle lying around, you could change it out for a clogged one, allowing you to continue printing while you work on cleaning the first nozzle.
- **To have multiple nozzle sizes.** The default opening size of the stock nozzles that ship with most consumer printers is 0.4 mm. This is a good general nozzle size, but you might want a different size for a variety of reasons. A larger nozzle opening would be good for printing larger parts and could print a lot faster than smaller nozzles; a smaller nozzle opening would be good for precise details but will usually increase the printing time. You may find it useful to keep a few different sizes on hand that you can swap out based on the project.
- **To upgrade to a better nozzle material**. Most stock nozzles are made of brass, which is good enough for basic applications. However, some filaments, like carbon fiber filament, have abrasive materials in them that are harder than brass and can damage the nozzle as they pass through it. In time, the opening of your nozzle can actually get larger as the inside is sanded

down by these abrasive materials. And then none of your projects will work out the way you hoped. For this reason, it can be a good idea to have nozzles made of different materials on hand, such as ones made of stainless or hardened steel. Be aware that different metals may heat up differently than brass, so keep that in mind when you're using a new nozzle.

- **Simply to have a better nozzle!** As is often the case with electronics, what comes by default in the package isn't always as nice as what you could get elsewhere. If you've chosen a budget printer, you might get budget parts and accessories, right? So if the manufacturing and quality of the stock nozzle aren't up to par, upgrading to a better nozzle can be an easy, quick way to improve the quality of your prints.

Filament storage

So here's something that might not have occurred to you: the moisture content of your printing material matters. If there's moisture in the filament, then when it gets to the printhead and gets heated up so it can be extruded, that moisture can vaporize and cause problems with the print, even causing it to fail.

Okay, you're saying, but what's the big deal? I'll keep my printing material out of the rain.

Unfortunately, it's not that simple: a lot of filaments actually absorb moisture from the air around them, meaning it's not enough just to keep your filaments in a safe, dry spot.

So what do you do? There are a couple of different ways you can keep your printing material dry.

The cheapest, easiest way is to store it with a desiccant. You know those silica gel packets you sometimes get in packages, the ones they always warn you about not eating? Well, you can buy those in bulk and keep them with your filaments in a sealed container. Some people also dry their filaments out in a low-temperature oven.

Want something a little more high-tech and a little less DIY? There are some great options available for purchase that will keep your printing materials dry and ready to go. For instance, you can get a filament dryer or dry box: a storage box that heats up enough to dry out your printing materials. You can even get ones that you can store your filaments in while you're printing from them, and they'll make sure that no moisture is absorbed by your filament right up until the moment it enters the printhead. If you've got a filament that you're confident is currently dry, and you want to keep it that way, you can purchase vacuum-sealed storage containers that keep your printing materials so tightly locked up that no moisture can enter.

(If you don't feel like paying, but you are feeling a little crafty, you can find tutorials online showing how to create your own dry box out of things you can get from the hardware store.)

Smoothing and finishing

So here's the thing about 3D printers: they don't create perfect products. FDM 3D printers, in particular, create a very particular sort of product. Because they set the printing material down in layers, you'll often see the little ridges of the individual layers, especially on curved surfaces in the final product. They also can't always do the finest of details, and sometimes they leave behind unwanted bits of plastic or fail to make completely clean and clear holes and details.

Now, you can do things to combat that: careful creation of your 3D model and fiddling with settings, and using certain nozzles. But the fact remains that even the most carefully printed object will often keep those distinct layer lines or not have quite the finish or level of detail you wanted.

Maybe you don't mind, depending on what you're creating, but if you do, there are lots of options available for perfecting your object once the print is complete.

For smoothing the print, sanding it is a great place to start; it's often the first step before using some of the other options we're going to discuss here. But while this will help, if you really want to give your final product a smooth surface, consider some of the following products and accessories. (Keep in mind that the printing material you choose will affect your options for smoothing; not every product will work with every printing material.)

One option is to fill in the gaps using some kind of liquid or paste. Many people have a lot of luck polishing their prints, just like you would with a metal object; the polish will fill in spots, and then you can polish the whole surface. High-fill spray-on primer, like the ones you could get at any hardware store, can do it as well. If you want something that's designed especially for 3D printing, you can find coatings for purchase online that you brush onto the object, giving it a smooth, professional shine. If you do choose one of these options, make sure you've researched whether it works well with the particular material you chose to print this project. That last thing you want is to wreck something you've already printed.

Another option is heat: because many printing materials are thermoplastics, they will soften when heat is applied to them. A standard heat gun, like you'd use for removing paint or wallpaper, can be used to soften and smooth the surfaces of your printed project. This is one option that requires a lot of care and a delicate hand. However, if you're not careful, there is a very real chance of you ruining your project past the point of repair.

A third option is acetone. Now, it's important to note that this only works if you've printed with ABS, as it reacts to acetone in a way that, say, PLA does not. But if you have used ABS, this is a great option because the acetone will break down the outer layer, leaving a much smoother surface behind. For simple, small-scale smoothing, you could purchase a smoothing pen, such as the kind sold by Filabot; these pens have a tip that spreads acetone. This will allow you to control exactly how much smoothing you want and where you want it to go.

If you want to smooth the entire surface of an ABS print, acetone vapor smoothing could be the right choice for you: this involves enclosing the print in a sealed container with acetone vapor and allowing it to smooth the outermost layer of the print gently. This is something you can DIY with tissues and a sealed container, but if you see yourself wanting to do this process on many 3D prints, you might want to consider purchasing a dedicated smoother, of which there are a number available online. These products will seal your print uptight and then smooth it with a fine vapor of acetone, and the convenience and ease of such a product may make it worth your while.

(One interesting option to look into is the Polymaker Polysher, which gets your prints to an incredible level of polish. The catch is that you have to buy Polymaker's own special filament to get the desired effect. Based on what I've seen, though, the end results are pretty impressive; if you want to sell the things you print or display them in your home, it could be worth it!)

Note: Keep in mind that everything you smooth your print, you're either taking off the top layer or filling it in, so doing so can cause you to lose some of the very finest details of your object. The more you smooth, the more details will be lost.

So what about finishing? Well, for that, there are finishing tools and kits. On the simple side of things, there are kits you can buy that come with blades, carving knives, needles, brushes, and other tools that will make it easier to perform important finishing tasks like slicing off support structures, clearing out holes or fine details, and smoothing out edges and surfaces.

Want something with a little more oomph? You can also buy handheld finishing tools with different metal attachments that heat up. This is a great option if you see yourself doing a lot of post-print work to get your models perfect; they'll be a lot faster and easier to use than standard blades and brushes.

Using these products and accessories for smoothing and finishing prints can be a great way to make prints from a cheaper 3D printer look like they came off a professional machine. If you want to sell or display your 3D prints, but you don't want to spend thousands on a high-end machine, a few extra accessories and a little bit of elbow grease can turn an okay-looking print into a great-looking one.

OTHER ACCESSORIES

We've barely scratched the surface here on the types of upgrades and accessories you can get for your printer. Unfortunately, many of the more powerful and advanced upgrades also require quite a bit more work and understanding of the inner workings of your printer. Hence, as I mentioned earlier in this chapter, we're not going to go into those here. For more advanced upgrades and accessories, check out later books in our series!

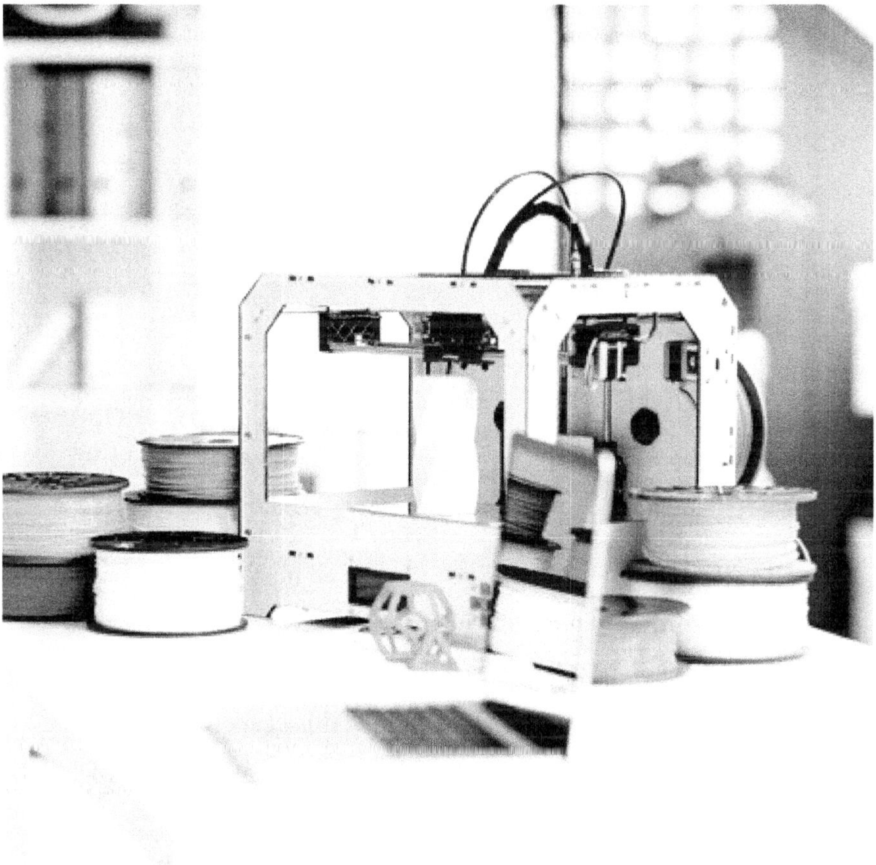

4

PICKING PRINTING MATERIALS

All right, so we've talked about printers and printer accessories. Now we need to talk about printing materials.

As I mentioned in chapter 1, you can use all manner of materials for 3D printing. There are people printing with materials as varied as concrete, plant-based meat slurry, and living human cells. What we're going to talk about in this chapter, though, is the sort of materials you'd be likely to use with a normal consumer 3D printer.

Most printers will come with some test printing materials in the box, but that's not going to last you long. If you want to hit the ground running with your printing, it'd be wise to buy some material around the same time you buy your printer.

PICKING PRINTING MATERIALS | 71

3D printing materials are also sometimes referred to as "filament," as you've probably noticed me doing throughout this book. This is because the material is often in the form of a long continuous thread, or filament, wound onto a spool.

(It's also possible to print from pellets, but as you can imagine, that requires some different printing equipment, and there aren't all that many consumer 3D printers that support it. It also doesn't really offer any advantages over filaments, at least not that I've found, so it's probably not going to become a popular print method any time soon. But I thought I'd bring it up in case you saw pictures of pellets online and were curious and confused about it like I was the first time I saw it.)

In this chapter, we'll talk about the considerations that you ought to keep in mind as you're picking a 3D printing material; then, we'll talk about some of the materials that are out there to choose from.

Note: Be aware that I'm just going to go over the most popular and basic materials; there are more types of materials available than I can mention in the space we have. Also, be aware that I might talk about a particular material. Still, there are dozens of different products available by different names that all fall under the umbrella of this material. For instance, we're going to talk about PETT, but if you buy any, you'll most likely buy it under the brand name "t-glase." So if you're searching for one of these materials online, don't be surprised if you see someone selling it under a different name, and only if you look close can you tell what precisely they're selling.

Also, be aware that certain brands of printers might insist that you need to buy their proprietary brand of filament; Cubify used to do so with their CubePro printers, although they've since shut down their 3D printing business. While this is rare, it's worth double-checking your printer doesn't have such requirements.

THINGS TO KEEP IN MIND

The very thing that makes 3D printing exciting—the flexibility, the customizability, the ability to do nearly anything you want—can also make it pretty overwhelming to get started. You've probably heard dozens of acronyms: PLA, PVA, PET, PETT, PETG—I mean, how are you supposed to keep these all straight and know which to use?

Well, I hope that by the time we get through this chapter, you have a better understanding of all this. And the best place to start is to look at what you have and what you need.

What printer do you have?

The first thing to keep in mind is that not every printer will work with every type of printing material. Certain materials have specific requirements that certain printers can't meet . . .

and it probably won't surprise you to hear that the more affordable the printer, the less it may be able to handle.

Your two biggest concerns have to do with heat. Certain filaments need to be heated up to certain temperatures to reach the point where they can be extruded properly, and certain filaments also fare better if they're extruded onto a heated print bed. Now, most printers will meet those two criteria, but some of the most basic, budget printers won't; the heated print bed, in particular, seems to be one feature that often gets chopped when companies are trying to figure out how to create a budget printer. So make sure your filament and your printer are compatible before you start.

Another thing to keep in mind is that, as I mentioned in the previous chapter, certain materials can damage the stock brass nozzles that come with most consumer 3D printers. Thus, you'll need to upgrade to a hardened nozzle (or even a fancy ruby-tipped one) if you want to print with those. So if you don't already have a hardened nozzle and want to use one of these abrasive materials, you have to decide if it's really worth it to you to pay for the new nozzle.

What are you printing?

Different materials are going to be more or less appropriate for different applications. We're going to talk about materials that are food-safe, that glow in the dark, and that are flexible, each of which would be great in certain applications and less so in others. Which material you need is going to depend on what you intend to do with the final product.

Note that some of the fancier materials are going to cost you more than the basic materials would, so make sure you've factored that into your plans.

What size filament does your 3D printer need?

Printing materials tend to come in two sizes: **1.75 mm** and **3 mm**, which refers to the diameter of the filament. Your printer documentation should tell you which size to buy for your printer.

The two are honestly pretty well-matched in terms of which one is "better." 3 mm filament can be easier to use and causes fewer clogs. However, these days, printers that use 1.75 mm are more popular, meaning more manufacturers are producing more materials that use 1.75, meaning you have more options. Some of the more obscure materials might be hard to find in 3 mm. So if I were buying a new printer right now, I'd probably lean toward buying one that accepts 1.75 mm filament.

Got all that? All right, let's move on to some of the printing materials that are out there.

BASIC PRINTING MATERIALS

Now, when I call these materials "basic," I don't necessarily mean that they're low-quality or even that they're the easiest to use (though they do tend to be on the easier end of things). What I mean is that these are the materials that I think of as being appropriate for basic printing needs, not for specialty prints. If I need to quickly print a little widget to sit on the edge of my desk and keep all my charging cables organized, I'm going to reach for one of these materials.

PLA

Definition: *Polylactic acid, or PLA,* is a polymer made from biological materials; these materials make it biodegradable. It produces durable, opaque prints.

About: This material is very popular with beginners and with budget 3D printers. It's fairly easy to use, and notably, it's pretty sticky. This means that, unlike some other materials, it doesn't require the use of a heated print bed in order to make the first layers stick and print correctly, which is great if you have a budget 3D printer that doesn't have a heated platform. It also melts around 350° F (180° C), which is pretty standard—any consumer printer you buy will be able to handle that temperature.

You'll be seeing PLA mentioned a lot on this list because a lot of composite materials—materials where bits of a material like wood or metal are mixed with a base—PLA is used as the base material.

Uses: This is great, as I said, as a basic material to get started doing projects with. It's food-safe, in its pure form, so it's great for things like cookie cutters and short-term food containers, like water bottles. (Check the label to be sure the brand of PLA filament you're buying doesn't have any additives that would keep it from being food-safe.)

As a side note, one use that's pretty interesting but that I imagine most of you reading this book won't be doing is in medical applications. PLA is biodegradable, as I mentioned; over time, it breaks down into non-toxic lactic acid (in certain conditions—don't worry that you're going to print something, and one day it'll just vanish). This makes it a great option for things like screws, pins, and meshes that can be placed in a patient's body for medical purposes and then left there: within a year or two, it'll break down into lactic acid!

Pros: As I said, the fact that you don't need a heated print bed makes this a great option for budget 3D printers. It's also easy to use—it's pretty forgiving if you don't heat and cool it exactly right—and produces sturdy builds.

My personal favorite thing about it? It's environmentally friendly, which isn't something you can say about a lot of plastics. It's made of lactic acid building blocks, making it sustainable and renewable. The process of producing it produces fewer emissions than a lot of other materials. And it's biodegradable, as I said above; under the right conditions, it can break down fairly quickly. So if you ever get to feeling guilty about all the plastic being produced by the 3D printing industry and the world at large, using PLA might make you feel a little bit better.

Cons: PLA doesn't produce prints that are quite as high-quality as certain other materials you'll find. It can also be brittle and therefore not very impact-resistant.

One big thing to look out for is the fact that, like all thermoplastics, PLA can be heated, then cooled and set, then reheated. Unfortunately, it doesn't need much heat for PLA to start to soften; prints made of PLA start to deform at temperatures as low as 140° F (60° C). I heard a story about a friend of a friend who printed a full-scale replica of R2D2 using PLA for a lot of the frame, and one time had to leave it out in his car on a hot day. When he got back to the car, poor Artoo was looking worse than that time he was nearly eaten in the swamps of Dagobah.

ABS

Definition: *Acrylonitrile butadiene styrene, or ABS,* is a water- and chemical-resistant plastic that produces tough, sturdy prints.

About: ABS has long been a popular material for 3D printers because of its low cost and the fact that it creates durable final products. (It can, however, be susceptible to long-term UV radiation, so this might not be the absolute best choice for something that's going to spend a long time in the sun.)

A fun fact about ABS: this plastic is what Legos are made of. As we talk about its benefits below, you'll see why it's an excellent choice for these toys.

Uses: As I said, it's durable, which makes it great for toys like Legos. That durability also makes it an excellent choice for things that are going to take some hard-wearing, such as automotive parts or gears. It also shows up in some surprising places: Kawai uses ABS in its pianos, and Yamaha uses it to make those recorders you may have had to learn to play in the 3rd grade.

Pros: The low cost of ABS makes it a great choice for beginners who may go through a lot of filament as they're learning how to use their printers. It's also tough and durable, as anyone who's stepped on a Lego can attest to.

It's fairly easy to print with, and it's much more heat-resistant than PLA; your ABS prints are probably not going to deform if you leave them in a hot car. And they don't react with water or many household substances they may come in contact with.

One characteristic that makes it excellent for Legos and other toys is that it takes color really well; mixing in pigment doesn't have adverse effects on the material. So you can buy fun, bright colors for action figures and toys.

Finally, unlike PLA, ABS is not biodegradable. However, it can generally be recycled. In fact, you can find stores online that sell recycled ABS filament (rABS). If you want to help reduce the amount of ABS waste out there, that could be a great option!

Cons: ABS has a higher temperature requirement than PLA: you need an extruder that can get up to around 425° F or 220° C. Most printers can handle that, but you'll want to double-check.

One thing to be aware of is that ABS can warp and contract as it cools, so you'll want to control its cooling. As I mentioned in the chapter on accessories, an enclosure (either purchased or DIY) will help to control the dropping of the temperature around your final product. Also, a heated print bed can keep the lowest layers from warping before the upper layers are laid down.

Speaking of heated print beds, you're definitely going to want one for an ABS print because ABS sticks much better to a heated bed. Many printers have a heated print bed—it's generally only the budget printers that don't—but if yours doesn't, you'll need to find ways to increase bed adhesion (which we've talked about in a few other chapters).

Make sure you print in a well-ventilated area because ABS has an odor when it's heated up; it's not the most pleasant thing in the world, and in high enough concentrations could possibly become harmful.

Nylon

Definition: *Nylon* is a synthetic polymer made of polyamides (a protein found in silk and wool).

About: Though nylon is relatively new to the world of 3D printing, it's been around as a material for quite a while: it was developed by DuPont in the 1930s and is the first commercially successful synthetic thermoplastic polymer. And yes, in case you're wondering, this is the same material that women's nylon stockings are made of. (An interesting fact: nylon stockings were sold starting in 1940, but supply almost immediately ran low because nylon was needed for parachutes during World War II.)

It won't feel like stockings or parachutes when you're printing with it, though! You'll find prints made of nylon filament to be tough, durable, and resistant to damage or abrasion. In the short while it's been used in 3D printing applications, it's quickly become a very popular material.

Uses: You mean other than stockings and parachutes? Nylon's durability makes it great for things that are going to take some wear and use, like gears, screws, and bolts; its strength makes it great for cable ties, mechanical components, and tools.

Pros: Even with the strength and durability I talked about above, nylon can be somewhat flexible if printed thin enough.

It's less brittle than PLA or ABS, with high impact resistance. And it's lightweight! It's also relatively low-cost, making it a great option if you need to print large parts.

It's non-toxic, and it can produce nice smooth prints. Note that brands vary on whether they emit unpleasant odors during printing; check the label and read the reviews of the brand you buy.

Cons: Nylon does have some more finicky characteristics and requirements. Some varieties require quite a high temperature for printing: up to 250° C and more, which is higher than some extruders can accomplish. So you may need to upgrade your extruder or even get an all-metal hotend to print with nylon. (There are lower-temperature nylon products available, though, so keep an eye out for those.)

Like ABS, nylon also requires a heated print bed to get the first layers down correctly, and like ABS, it is prone to warping as it cools, so you may want to use an enclosure as you print you can control the temperature around your print.

One uniquely difficult thing about nylon is that it's pretty highly hygroscopic, which means that it absorbs moisture from its surroundings; I've heard that it can absorb up to 10% of its weight in moisture in a single day. As I mentioned in the chapter on accessories, moisture in your filament can cause major problems when you go to print. The moisture is turned to steam as the filament is heated up, leading to all sorts of finish and structural problems as you print. So if you use nylon, you need to really make sure you're keeping it dry. Refer back to the chapter on accessories for some hints on how to do so.

Nylon is not biodegradable (depressingly, it accounts for 10% of the debris in the ocean), and while it can technically be recycled, it's hard to find places that do it. So be thoughtful about how you use it.

PET/PETG/PETT/t-glase

Definition: *Polyethylene terephthalate, or PET, is a thermoplastic polymer resin in the polyester family. In 3D printing, variants called PETG (polyethylene terephthalate glycol-modified) and PETT (polyethylene cotrimethylene terephthalate, sold commercially as t-glase) are popular.*

About: PET was first patented in 1941 and has a variety of uses outside 3D printing: it's commonly used in water bottles and packaging for food, and it's the same polyester that's used in clothing and cheap Halloween costumes (in fact, that's what the majority of this material is used for). When it's processed into filament for printing, however, it becomes sturdy and resilient.

One of its most unique properties is that it is translucent and can even almost look transparent, depending on how it's processed and shaped. So if you're after that see-through look, this is the printing material for you!

PETG is PET that's been modified with glycol; this makes it less brittle, more clear, and easier to use. PETT lacks the glycol and so is slightly more rigid than PETG but has better optical qualities. You can basically only get PETT from Taulman right now; they sell it under the name t-glase.

Uses: PET is considered food safe by the FDA, so this is a great option for cups, bottles, and other dishes and food packaging. Because it's waterproof, it's also a great option for things like vases.

Its unique appearance also makes it suited for translucent prints. But keep in mind that it's not going to be like looking through a glass; it's tricky to get absolute tons of translucency from this material, so even in the best-case scenario, it's not going to be crystal clear.

Pros: The big benefit here, obviously, is the appearance; not every plastic offers that unique appearance. It also creates a smooth and glossy finish. The filament can be colored without losing its see-through quality, so you'll find colored varieties for sale.

Prints made from PET are strong and have great impact resistance.

One thing that's great about PET and its variations is that they don't tend to warp much, and they adhere well to the print bed, meaning that heated beds and enclosures aren't required, though you may find them useful.

PET is not biodegradable, but it is recyclable. In fact, if you want to make the world a cleaner place, you can find shops online that sell PET filament made from recycled materials like water bottles.

Cons: Like nylon, PET can absorb a lot of moisture from the air, which will lower the quality of your prints. Dry your filament before printing using a dedicated filament dryer and storage system, or go full DIY with your oven and some silica gel packets in an airtight container.

Though PET prints are strong, they have a somewhat softer surface than many of the other materials we've looked at, making them a little more prone to wear.

These materials require a temperature of around 230° C, which is pretty high. If you're thinking of using one of these printing materials, I'd recommend you read up on whether your printer's documentation claims to be able to support PET materials. Check online to see what other users of your printer model have to say on the subject.

Finally, don't pin all your hopes on your final product being transparent. It'll definitely be much more transparent than something printed from ABS, but it's not going to be like the final product was carved from crystal. Even with t-glase, where the big selling point is the optical quality of the material, the most the manufacturer claims is that it's "considered colorless per industrial classifications." Get online and look at pictures of

PET prints to get a sense of what you can actually expect if you go for one of these materials.

ASA

Definition: *Acrylic styrene acrylonitrile, or ASA,* was developed as an alternative to ABS that would be more suitable for outdoor uses because of its higher UV resistance.

About: ASA is chemically similar in a lot of ways to ABS, as you might guess from looking at the names: acrylic styrene acrylonitrile and acrylonitrile butadiene styrene. The big difference is that ASA incorporates acrylate rubber, while ABS uses butadiene rubber; this gives ABS a number of advantages over ABS, such as the aforementioned resistance to ultraviolet radiation.

Work on developing ASA started in the 1960s, and the material has grown a lot in popularity since then. In addition to being popular in the 3D printing world, it's used a lot in applications where the final product will be exposed to the weather, like cars or lawn and garden equipment.

Uses: These characteristics I've been mentioning make this material a great choice for applications where it'll be used outside; if you're going to print yourself a lawn gnome or a custom mailbox, this will be the plastic to use. It's also great as an all-purpose material, given its similarities to ABS, though its higher price point may deter you from using it for applications where you don't specifically need the UV resistance.

Pros: The different formulation gives ASA a lot of advantages over ABS. As I mentioned, it has higher UV resistance; when left outside for long periods of time, it's less prone to yellowing and maintains its appearance and glossiness better than many other plastics would be in the same circumstances.

The material produces pretty toughly, sturdy prints, which further qualifies it for outdoor use. It also has better chemical and heat resistance than ABS and will weather better over time.

Cons: Though designed as an alternative to it, ASA shows a lot of the same weaknesses as ABS, especially where printing is concerned: it requires high temperatures, you really need a heated print bed, and it warps easily. It can also release some pretty strong (even potentially dangerous) fumes, so you'll definitely want to ventilate the area well when you print. It can also be a somewhat expensive filament to buy, so if you're not specifically printing for an outdoor application, this might not be the most cost-efficient way to print.

DISSOLVABLE PRINTING MATERIALS

You might be looking at that heading and thinking, "Why in the world would I want my printing materials to dissolve? Why would I want to print something that could vanish if I got it in contact with the wrong substances?"

Well, generally, you don't make final products out of dissolvable materials; what they're useful for is support structures. I mentioned this a little in chapter 2, but if you don't mind me repeating myself: one tricky thing about 3D printing is how to print pieces of your model with overhangs, where there's nothing underneath them to support them. Imagine trying to print a model of the Golden Gate Bridge, which is full of pieces that span gaps or extend out into nothing. How can you possibly print those properly? Each layer needs to sit on *something*, after all (unless the distance it's bridging is small or the overhanging part is jutting out at an angle of less than about forty-five degrees).

The trick is to include structures on your model that will support these overhanging bits. The hard part is that when the whole thing has cooled down, you have to cut or break the supports off, which is time-consuming and can damage your print if you're not careful. If you have a dual-extruder printer, however, you have a second option: have one extruder use the main material to print your object and have the other extruder use a dissolvable material to print the support structure. Then, when the print is done, you just submerge your product in whatever will dissolve the support structures. There's no cutting, no smoothing, no trying to hide the fact that something used to be connected there.

All of this is to say that this section is mostly useful if you have or are considering buying a dual-extruder printer. Still, even if you don't, keep reading: you may think of a creative use for these dissolvable materials.

PVA

Definition: *Polyvinyl alcohol, or PVA,* is a biodegradable synthetic polymer that dissolves in water.

About: PVA was first discovered in 1924, but it took until the 1950s for it to start being used much commercially. It's used in all sorts of applications, like making paper and mortar. Its biocompatibility has led to its use in certain medical applications, like soft contact lenses and artificial cartilage.

Because it's water-soluble and non-toxic (as long as it's in reasonably small quantities), it's useful for applications like bait bags. Fishermen buy PVA bags, put bait in them, and drop them in the water; the bag dissolves, releasing the bait and attracting fish (and all without introducing plastic waste into the environment). There's even been research into using them for time-release capsules for medication.

Uses: In the 3D printing world, PVA is basically only used for support structures for prints made of more durable materials. It's especially useful for really complex prints, with support structures in nooks and crannies, which you're going to have a hard time getting out with just a razor blade. All you do is submerge your finished product in warm water, and the PVA will dissolve away, leaving your beautiful final product behind. (Note that you'll be left with water filled with a sticky residue, so you'll want to be careful how you dispose of it.)

Of course, all this means that you need a dual-extruder printer: one extruder printing with your main material and the other extruder printing with PVA. If you don't have a dual-extruder printer, you'll have to make the whole print out of the same material and just get handy with the finishing tools when the product's done.

Though we've talked chiefly about support structure uses for this product, you may think of other applications where its dissolvability is a feature, not a bug—like the PVA bait bags I mentioned earlier.

Pros: We just talked at length about the obvious benefit: the ease of removing support structures made of PVA. But that's not all it has going for it: PVA is really easy to print with, not requiring high temperatures or an enclosure or a heated print bed. You won't need to modify your printer at all to use it.

This plastic is fairly biodegradable; this doesn't mean you should feel fine just tossing it in the grass at your local city park, but it may make you feel a bit better if you worry about plastic waste.

Cons: The very thing that makes this so useful—that it dissolves in water—can make it tricky to store. you need to keep it as dry as possible and away from water sources and even high humidity. Store it in an airtight container at all times.

It's also on the expensive side, so if you're doing a print with PVA as the support, you'll want to make sure your model is optimized to use as little support as possible.

HIPS

Definition: *High-impact polystyrene (HIPS)* is a hydrocarbon polymer popular in the 3D printing world for being dissolvable in limonene.

About: Polystyrene, first discovered in 1839, is one of the most commonly used plastics in the world: you'll find it in CD cases, plastic cutlery, containers, and more. Its foam form is very popular as a lightweight packing material; you might know it by its commercial name, styrofoam.

What we're talking about in particular, though, is high-impact polystyrene, which, as the name would suggest, is meant to be more resilient and absorb impact better than other types of polystyrene.

Perhaps its most notable feature with regard to 3D printing is that it dissolves in limonene, a liquid hydrocarbon best known as the oil in citrus peels (which is where the name comes from: the French *limon*, meaning lemon). Unlike the water needed to dissolve PVA, you almost certainly don't have access to a lot of limonene. So if you decide to use HIPS and dissolve it, you'll also need to buy a jug of limonene.

Uses: HIPS is low-cost, impact-resistant, and easy to fabricate with, so in the general manufacturing world, it's commonly used for all sorts of things: toys, drinking cups, signs, kitchen utensils, ceiling tiles, components in electronic appliances, and more.

In the 3D printing world, it's most commonly used for support structures, but it's also a great alternative to ABS for normal printing, being more dimensionally stable and lightweight than ABS. Yes, it'll dissolve in limonene, but how often does your average object come into contact with the oil in citrus peels?

Note that, as with PVA, using this to add dissolvable support structures to your print will require a dual-extruder 3D printer.

Pros: Obviously, the dissolvability is the real selling point to many people who do 3D printing; it makes removing support structures easy (and citrus-scented).

It's got other pros, though: it's low-cost, lightweight, and produces sturdy, impact-resistant prints. It's fairly easy to print with, with great dimensional stability.

If you're using HIPS as a normal printing material—not just one for support structures—you'll be glad to hear it's pretty easy to finish: once it's been printed, it can be sanded, smoothed, painted, and glued with relative ease.

Cons: This stuff really doesn't like adhering to the print bed; a heated platform is necessary to get it to adhere at all, and you still might need to prep the platform first with tape, glue stick, etc. It also prints at a higher temperature than a lot of other materials. Make sure your printer and your hotend can handle the temperatures needed. Your printing process will also benefit from an enclosure so the print can happen in a heated environment.

Keep in mind that if you use HIPS as a support material for an ABS print, and you leave the final product in limonene for a long time to dissolve the HIPS, the limonene can start to affect the ABS after a while. So keep an eye on it and don't leave it in there too long.

HIPS is very recyclable, but it's not always easy to find a recycling plant that accepts it; unfortunately, it's not very biodegradable. (I mean, just think about the amount of polystyrene plastic and styrofoam litter that's out there. It's a lot.) So keep that in mind when you choose to print with HIPS.

COMPOSITE PRINTING MATERIALS

So far, all the materials we've talked about have created plastic-looking prints. And a lot of items, that's exactly what you want.

But what if you wanted to use your 3D printing to create something that looked a little more impressive? A little more sturdy? A little more classy?

That's where composite printing materials come in. As the name suggests, they're a composite of two materials: a base polymer, often PLA, but sometimes others, in which are mixed particles or fibers of other substances. The thermoplastic base makes it printable, and the particles or fibers give it some of the properties of the other material.

Some composite materials increase the strength of the final product. Others are purely used for aesthetic reasons (and can even decrease the strength or durability of the final product).

Carbon fiber

Definition: *Carbon fiber filament* has short strands of carbon fiber in a base filament such as ABS, nylon, or PLA.

About: Most of the filaments I'll mention in this section are basically just used for aesthetic purposes. Carbon fiber filament, however, actually improves some of the mechanical properties of the base filament.

You've probably heard of carbon fibers, which are stiff, lightweight, and strong, with high tolerance to temperature and resistance to chemicals. This has long made them popular for engineering, sports, and military applications. You've no doubt heard of high-performance carbon fiber bike frames, laptop cases, race cars, and airplanes.

Carbon fiber filament isn't quite as impressive because it's only short fibers suspended in a thermoplastic. Still, it's a pretty extraordinary filament.

Uses: Its strength and low weight make carbon fiber filament a great choice for prototyping parts that need to be strong enough to work properly in demanding applications. It's also great for protective cases and other applications where high durability is needed.

Pros: As I said, a filament with carbon fiber in it is going to be stronger and more durable than a filament that's simply made of the base plastic (though not as strong and durable as a carbon fiber reinforced polymer, the type that would be used in a Formula One race car). Prints made from this filament are lightweight, especially when compared to their strength. And the carbon fiber bits in it give it good dimensional stability.

Like most composite materials, this filament takes on a lot of the printing characteristics of whatever its base is. So if you've chosen one with an easy-to-print base like ABS, PLA, or nylon, it will also be pretty easy to print with.

Cons: As you might imagine, this filament can be pretty hard on your printer; the carbon fiber bits in it are rather abrasive. In fact, if you're using the basic stock nozzle that came with your printer, a carbon fiber filament can damage it to the point of being unusable in the space of only a few prints. It can wear down the nozzle until the hole is too big, causing printing to become sloppy and imprecise. So if you're going to start using carbon fiber filament, you definitely need to get an upgraded hardened nozzle that can stand up to the abuse.

If you ever thought you might want to print extensively in carbon fiber—maybe if you're using it to fabricate lots of parts and prototypes—you can buy special dedicated carbon fiber 3D printers. It may not surprise you to learn that they are not cheap. While we're on the subject: carbon fiber filaments are also often not affordable.

Wood

Definition: Wood filament features wood particles in a base polymer, usually PLA; the filament tends to be about 30% wood particles and 70% plastic.

About: So basically, this is the complete opposite of carbon fiber filament: you choose it not for its mechanical properties (in fact, it can actually worsen the mechanical properties of the base plastic) but purely for how it looks. Not interested in learning how to whittle? Don't worry; the wooden toy boat of your dreams is only a spool of wood filament away.

You can get filament with a variety of different wood particles—mahogany, bamboo, ebony, even cork—for different final looks. (Some companies sell wood-colored filament with no real wood particles in it, so keep an eye out for that.) You can even get fancy and have your hotend switch to different temperatures throughout the prin. Certain wood filaments get darker and higher temperatures, so you could deliberately darken the final product in certain points to mimic the variation in color found in real wood.

Be aware that for best results, you'll probably want to sand the wood when you're done to get the right look and feel.

Uses: This is great for when you want the appearance of real wood! Use it for toys, props, sculptures, and decorations: print a fake bamboo toothbrush holder or a tiny ship to stick in a bottle.

Pros: One interesting thing about printing with wood vs. shaping wood is that shaping wastes more wood: you start with a block of wood and then cut and carve at it until it's the shape you want, and you discard everything you just took off. And since PLA is among the more environmentally friendly materials we've discussed here, a wood filament with a PLA base is something you can feel better about using.

This filament is easy to use because it maintains most of the characteristics that make PLA easy to use; it doesn't require high temperatures, a heated print bed, an enclosure, or anything.

Many prints suffer from the fact that when you look at the finished product, you can really see the separation between layers. With wood filament, however, that can actually add to the final look! Especially if, as discussed, you've fiddled with the heat level to get that more organic look.

Once finished, the print can be sanded, lacquered, stained, etc. Once you've done all that, it looks pretty convincing!

Cons: One downside is that you will definitely want to do some sanding, lacquering, staining, etc., to get the final print looking its best.

Another downside is that adding the wood particles to the PLA makes the whole filament a little more brittle; if the filament has to round any sharp corners on the way from the spool to the printhead, you could see breakage. And speaking of breaking, remember that the prints you create are made of a polymer mixed with wood fibers, not from the heart of a mighty oak tree: they'll be sturdy but not as strong as you might find in real wood. In fact, the addition of wood fibers to the PLA actually lessens that material's impact resistance, making wooden 3D prints somewhat brittle.

Metal

Definition: Metal filament features metal powder in a base plastic, generally PLA.

About: A lot of what I'm going to say here echoes the wood filament section because, once again, these are particles of another material suspended in plastic. In this case, it's a variety of metal powders: you can get bronze, copper, brass, and more, depending on the look you're after.

And honestly? The prints look pretty good, though you might not think so when you first see the final product; generally, you'll have to sand and polish up the print to really get it to look right. One interesting characteristic of these prints is that because they have metal powder in them, they have more heft to them than your standard plastic print; when you lift them, you can tell they're (partially) metal.

Uses: This filament is great for statues and decorations that need that metallic look; imagine creating a scale replica of the Eiffel Tower in a material that has the look and heft of real metal. This is also a great choice for jewelry, costume pieces, and props.

Pros: Most of us don't have the capacity to fabricate things out of metal at our kitchen tables; this is a nice compromise that's much more accessible to the average person than industrial metalworking machines. As I mentioned, these metal prints have a nice, convincing weight and look to them (after they're polished and they've got that certain shine). Do be aware that some companies sell filaments that are metal-colored but actually contain no metal; make sure you read the label and know what you're getting.

Because PLA is usually used as the base, this filament isn't overly hard to print with; it's going to have the same temperature, and print bed needs as a standard PLA filament.

Cons: As with wood filament, you have to do some work at the end to get metal filament prints looking their best. And as with wood filament, the presence of foreign particles inside the filament can make it more brittle and liable to break if it's forced to go around tight corners or fold back on itself.

Because it's heavier than the average filament, the metal filament has some trouble with overhangs wanting to sag; you may need even more support structures than you would use with another filament.

The metal powder within the filament is somewhat abrasive and can damage your nozzle; if you're using a basic stock nozzle, you'll need to upgrade to a hardened one.

And yet, none of that weight and abrasiveness means that this filament is strong and durable as metal; printed parts are actually pretty brittle.

Stone

Definition: *Stone filament* features powdered stone or chalk in a polymer base (usually PLA). It creates prints that have the look of carved stone.

About: This'll sound familiar from the wood and metal filaments. this is a basic plastic with powdered stone in it. It's a bit more unusual than the other two, but I brought it up because I find it really interesting that this is something you can do. You can find it in different colors, including some with multiple colors to give it that more realistic feel.

That being said, if I'm perfectly honest, I tend to find this to be the least convincing of these types of filament (stone, metal, wood) that attempts to mimic another substance. I think the wood and metal ones can look pretty great once they've been finished; I've yet to see the stone 3D print that I've thought looks convincingly like real stone. If you find one, let me know that I should have a little more trust in the stone filament.

Uses: This is largely going to be for decorative use: use it to create busts, chopstick rests, and replicas of the Easter Island heads.

Pros: As with the other two, this filament is usually reasonably easy to print with because the base is usually PLA or another common, easy-to-use filament: there's no special temperature or heated print bed requirements. It can be used to create something that most of us just don't have the chiseling skills to create on our own.

Cons: Unfortunately, as with metal, the addition of the stone powder lessens the durability of the underlying PLA; prints made from the material are often brittle and break easily.

Also, because the stone powder can have little bits in it, it can be somewhat abrasive; over time, it can damage your nozzle. If you're going to print with stone, you probably ought to upgrade to a hardened steel nozzle first.

This is definitely one of the less known and less common types of materials; when I Googled it, only a handful of purchasing options popped up. This isn't great for you as a buyer; more buying options mean companies are competing with each other on price, which is usually good for you. Still, you can definitely find some interesting options out there for stone composite filaments.

SPECIALTY PRINTING MATERIALS

This last section is about materials that fall into the "Other" category: materials with unusual properties that don't fall easily into the categories we've already mentioned. These are materials you probably won't use frequently, but that might be perfect for specific prints and applications.

Really, there are loads of specialty materials out there, with fun new ones appearing all the time; I don't have time to cover even a fraction of what's out there. So what I've chosen here are some of the filaments I personally find interesting and fun, but keep an eye out because there are lots of other great options out there.

Glow-in-the-dark

Definition: *Glow-in-the-dark filaments* involve glow-in-the-dark material added to a base, usually PLA.

About: We've all had glow-in-the-dark toys, right? Or at least glow-in-the-dark stars stuck to our ceilings when we were kids? This is the same sort of idea: if you expose the material to light, it glows for a while afterward. (As a note: with this kind of material, the kind of light you expose it to can affect how well and how long it glows. For instance, a good UV light up close is going to give you a better glow than if you left the object out on your kitchen counter and let it absorb the light from a distant incandescent bulb.)

Uses: This is great for kids' toys, Halloween decorations, costumes and props, and any other application where glowing in the dark would make it a little more fun. You might also find practical applications for this material; for instance, what if you used it to make household objects you often find yourself groping for in the dark, like light switches or plugs?

Pros: The base of this material tends to be PLA, which means that these filaments have all the same great benefits of printing with PLA: it's easy to use, and it doesn't require particularly high temperatures, heated print beds, enclosures, or special nozzles. Just print as you would with regular PLA, and you shouldn't have much trouble!

Cons: This is a lot of fun, and if you do some research online and pick a brand with really good reviews, you can expect a material that glows nicely in the dark. Just don't expect miracles. You've seen this kind of glow-in-the-dark plastic before, right? The glow usually doesn't last terribly long, and it's never going to be bright enough to read *War and Peace* by. Just keep your expectations reasonable, and you'll be happy.

Flexible

Definition: *Thermoplastic elastomers (TPE)* have a blend of rubber and harder plastic, making them sturdy but flexible. One of the most popular forms of TPE is thermoplastic polyurethane or TPU.

About: Thermoplastic elastomers have been available for a number of decades now. In the world of 3D printing, they add an interesting characteristic that most other materials can't: that prints can be deformed without staying that way long term.

Make sure you read up on the reviews and questions about the material you buy; different brands have different formulations, which means that some are super flexible and others are only a little bit flexible.

Uses: This flexibility would make it ideal for applications like phone cases, where a certain amount of bending is required, but you also want something sturdy to protect your phone. It's also great for toys—imagine using it for the tires of a toy car—and for vibration dampening.

Pros: The fact that you can flex prints made from this material set it apart quite a lot from many other plastics; this opens up a whole world of interesting possibilities for printing. It also makes fairly sturdy final prints, which are not much prone to wear and tear. This material has good material stability and thermal properties as well.

Cons: This is a material that is not easy to print with. People find it is prone to string—that is, as the printhead moves, it leaves behind long strings of filament strewn about the final project—and that it's not great at printing overlays. You'll want to design and optimize your model very carefully to avoid some of the worst problems related to retractions (when the printhead pulls back or retracts, some of the filament so that the printhead can travel to a new part of the model without leaving behind a trail of filament). These prints may also benefit from a slower print speed.

Also, the flexibility that is the focal point of everything I've just said about the material is also the cause of one of its challenges: it bends easily. So when a 3D printer is trying to force it into the printhead, the filament might not cooperate. If you think you want to print a lot with flexible printing materials, you might invest in a direct drive extruder. There are even companies that sell printheads specially designed to deal with flexible materials. If you're going to be using a lot of flexible filament, you might want to consider looking into it!

Really, just be aware that when printing with this material, you may need to experiment a lot with different settings before you figure out what's going to work best with your filament and your printer.

Conductive

Definition: *Conductive filaments* combine thermoplastic with a conductive substance to allow you to create prints that conduct electricity.

About: This is a relatively new product on the market, and it opens up a whole world of possibilities for 3D printing. The conductive element in these filaments is usually graphene, which is a carbon allotrope that conducts electricity well. Graphene itself is still fairly new; the potential applications are still being explored, but as new methods are developed that bring the cost of producing graphene down, we'll no doubt see it in even more places. And the filaments that incorporate graphene are the same way.

Uses: This is definitely the most specialized of all the materials we've discussed; I imagine 98% of the people reading this book will never find a use for this material. But if you like making or repairing electronic gadgets, this could open up a whole new world for you! Imagine custom designing electrical circuits or printing specialized styluses for use with touchscreen electronics. This material could also be used for creating capacitive sensors, like the trackpad you find on a laptop. All of these are definitely very specialized applications, but for a certain segment of the 3D printing population, this could be a game-changer.

Pros: Obviously, the first pro is that this plastic conducts electricity. Also, as with some of the other materials we've mentioned, because the base of this filament is PLA, in many respects, it prints like PLA; you don't necessarily need a heated print bed or an enclosure or anything.

Cons: The most important thing to keep in mind is that while this material is more conductive than most 3D printing materials, it's still less conductive than truly conductive materials like copper. It's really best suited for small, low-voltage applications, where it's only going to need to work with low-power devices. The slow conduction is definitely not suited to high-power applications.

This is also among the more expensive filaments, given the high costs associated with the use of graphene.

One problem that I've heard a lot of people report when they're using this material is that the addition of graphene makes the filament more brittle than pure PLA would be. This can affect your final prints, and it can also cause the filament itself to snap as it's traveling from the spool to the printhead. You'll want to make sure your setup is configured in such a way that the filament doesn't have to go around any tight corners, and you'll also want to be careful with how you design and handle your prints. Some people have had good luck printing a PLA casing to go around the outside of the conductive print. This adds some strength and durability to the overall final product.

This was a long chapter, but I hope you've found it useful! There are lots of great choices out there and no right or wrong answers: with each print you make, the right filament to use will depend on your printer, your project, and your desired final outcome.

One final word on filaments: assume that the first few prints you do won't turn out perfectly. With that in mind, start out using one of the cheaper filaments, like PLA or ABS, until you get the hang of things. Definitely don't start with cartoon fiber or conductive filament first: you'll probably end up wasting some money that way.

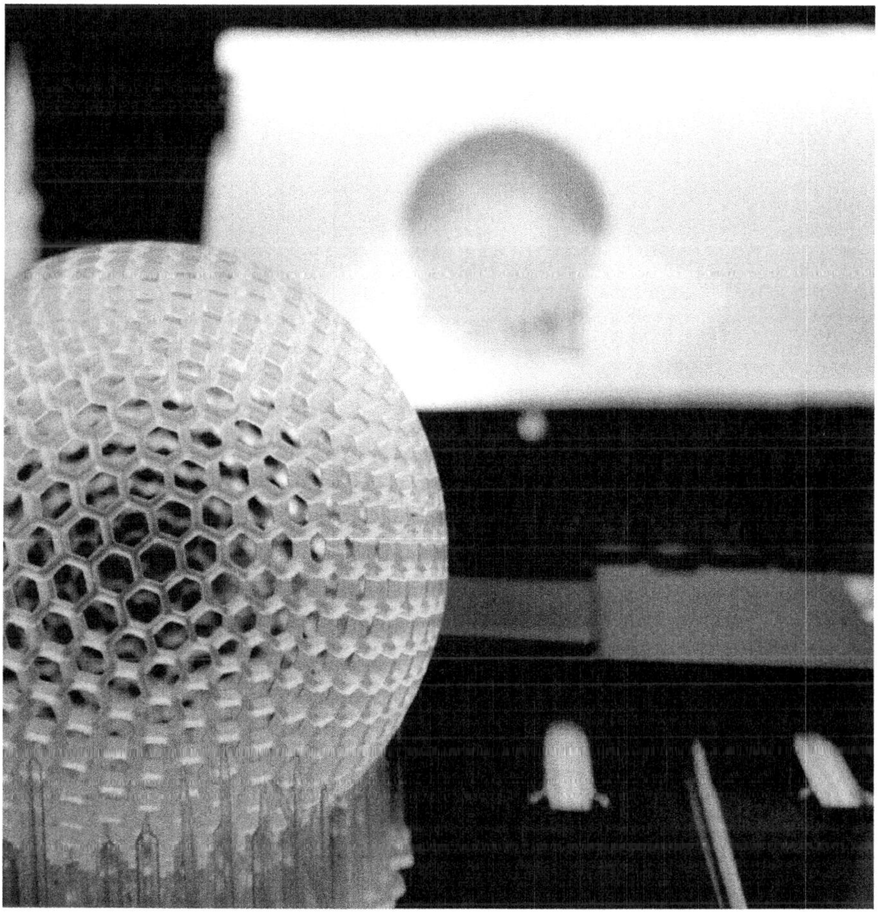

3D PRINTING SOFTWARE

So far, we've talked a lot about hardware and materials: what printer to choose and what material to use. But it's not all hardware; you're going to spend some time on your computer as well, working with 3D printing software. How deep you go down the software rabbit hole will vary, depending on your needs and interests. But at least some computer work is an integral part of every print.

So to start with, let's look at a basic overview of what you'll need to do software-wise as part of 3D printing.

1. Get your hands on a 3D model to print.
2. Prepare the model for printing.
3. Send the model to the 3D printer.

This is one of the exciting parts, right? You've chosen the printer and the material, you've done all the technical work of setting it up, and now you get to decide what you're going to print and how you're going to print it.

We're going to talk about each of these steps in detail. Keep in mind that this is a basic overview; many of the details will vary based on the printer and the software you're using, so I'm not going to get into a lot of specifics here. For specifics, you'll want to check the manuals and support documentation for your printer and software.

All right, let's dive into the process.

GET A 3D MODEL

I deliberately chose the word "get" here because that's what we're going to cover: getting your hands on a 3D model in order to print it. This book isn't going to cover 3D modeling; this is meant to be an overview and an introduction, and 3D modeling—which can be a pretty involved, complicated process, especially if your model and/or your software is complicated—is outside the scope of this book. There are other terrific sources out there to learn more about 3D modeling if that's something you'd like to get into.

A word about file formats

Nearly every piece of 3D modeling software out there is going to have some special proprietary format they save in, which is no good for sharing files or for printing; there's a good chance that your printer won't read whatever obscure format the software uses. It's important to have a common file format that just about every piece of software can export to, and every printer can accept. And that's where STL comes in.

Definition: *STL* is a file format for representing 3D objects. The STL stands for stereolithography, which is another type of 3D printing that uses a whole different mechanism from the FDM printing we've been talking about. The format was originally invented by the company 3D Systems, but it's become very widespread throughout the world of 3D printing.

Now, there are other formats out there that you may use from time to time. But STL is a great place for us to start the discussion because it's a widely used, widely compatible file format. Think of it as the lingua franca of 3D printing.

So as we talk in this section about finding 3D models, you'll want to make sure the models you find end in .stl. Of course, if you find other formats, you may be able to open them and export them as an STL, but that requires that you have the right software to open the file. It's much easier to find a .stl file, to begin with.

Finding models

So, where do you get a 3D model if you're not going to make it? You have a number of options here. To begin with, some printers will include basic models to use as a test print on your new printer; if all you want to do is print something to get a feel for the process or to test out your newly purchased printer, start there!

Next, go online to check out the millions of models that can be found on the Internet. Here are some great sources to check out.

Thingiverse is probably the best place to start, both for our discussion and your search for files. This is the largest collection of 3D models on the Internet—more than 2 million, last I heard —and it is all 100% free.

It's run by Makerbot, which is a name you should recognize from our discussion on the history of 3D printing; they showed the first consumer 3D printer at a tradeshow back in 2010. I quite like Thingiverse for a number of reasons, but one of them is the commitment to an open platform. Check out the About page on the website, and you'll see the following: "In the spirit of maintaining an open platform, all designs are encouraged to be licensed under a Creative Commons license, meaning that anyone can use or alter any design." So, in addition to the models being free, many of them are freely available to alter at will.

Thingiverse encourages anyone who wants to contribute to do so, whether they're a professional, an amateur, or a newbie, and there are good and bad things about this: on the one hand, this means that there are loads of models that people have contributed, so you're more likely to find what you're after.

On the other hand, you're going to find a wide range in terms of the quality of models; some will definitely be better than others. Luckily, the platform facilitates and encourages interaction: open up a model, and you'll be able to see the number of people who've liked it, comments from other users, remixes (models that people have made by modifying or borrowing from this model), and makes, which are pictures that people upload of prints that they've made using this model. So from looking at these likes, comments, and makes, you can usually get a pretty good sense of whether this is a good-quality model.

Basically, if I am looking for an STL file for 3D printing, this is the website that I check first.

Check it out at www.thingiverse.com.

CGTrader is the opposite end of the spectrum from Thingiverse. While Thingiverse is an open platform to freely trade models for 3D printing, CGTrader is a marketplace for buying and selling many types of 3D models, not only those that are suitable for 3D printing. It's still a great place to look for models for a couple of reasons. First, the models they offer are of very high quality as professional designers make them.

Second, though it is mainly for buying and selling models, it does have a sizable collection of free models.

CGTrader has been around since 2011; it was founded by a 3D designer and was conceived as a designer-friendly marketplace. Apparently, that was an idea that there was a big market because the site has grown by leaps and bounds since then; it now has 1 million models available and nearly 4 million registered users and includes quite a few Fortune 500 companies among its clients.

As I said, this means that these are high-quality models. Just make sure, when you're browsing what's on sale there, that you've selected "3D Print Models" when you go to search for 3D models. You can also tell from the model's page whether it's appropriate for 3D printing; it'll generally say so in the description or the details.

To find free models, select "Free 3D Models" when searching for 3D models. However, you may find it worth it to purchase a model if you find one that you like. They tend not to be terribly expensive—generally from $5 to $50—and you can find some fantastic work there. And if there's a design you love, but it's in the wrong format, some designers make available the option for you to reach out and request a format conversion.

This is also a great platform to find designers for custom work. If you find a model you like and you like the designer's work, you can use the Hire Me button to reach out to that person about hiring them for a custom job, or you can use the Freelance 3D Designers platform to post jobs and hire freelancers. If you just want to print a toothbrush cup for a bit of fun, you probably don't need to take this step. But under certain circumstances—maybe you intend to sell your prints, so any upfront costs would be made up by your sales—this might be the perfect way to get a model that's exactly what you want.

Check out all of this at www.cgtrader.com.

Cults is honestly not one of my favorites. It basically does what CGTrader does: it offers a mix of free and not-free models, mostly of high quality and not overly expensive. However, the collection is not as large as the one on CGTrader, and I find the website cluttered with ads and hard to use.

Okay, you're asking, then why even bring it up? There are a couple of reasons here: first is the fact that the website supports English, French and Spanish, so if you're more comfortable in French or Spanish than English, you're in luck!

Second is that Cult is trying really hard to be a more social experience for 3D printers; you can follow designers you like, which is fun if you want to see what new models they come up with. They also post a lot of contests.

So if you see yourself being interested in either of those two things, you might want to give Cults a try!

Check it out at www.cults3d.com.

MyMiniFactory is an interesting website. At its most basic, it's a marketplace for buying and selling 3D models; they may not be as professional as those at CGTrader, but they are often cheaper, so it's a trade-off. It also offers quite a few free models; unfortunately, I've never found a way to filter the models to show only the free ones. Fortunately, once you're looking through the available models, the ones you have to pay for are pretty clearly marked.

Now, it's not as slick as CGTrader, and the website is a little cluttered and not my absolute favorite to use; it also has a smaller collection. So why bring it up? Well, there are a few things that make MyMiniFactory stand out.

The first is that the website definitely leans toward gaming; there's a truly astonishing amount of models related to tabletop gaming on there. Don't get me wrong—there's loads of other stuff too, though even browsing a category like light fixtures brings up quite a few fantasy- and gaming-themed models. MyMiniFactory hosts information about crowdfunding campaigns related to gaming as well. So if that's something you're looking for, you'll definitely want to check this website out first.

Another thing that's nice about MyMiniFactory is that they claim that all models are run through a software check and then are test printed before being published, so you can rely on those files to be pretty reliable and useful.

If you're interested in the creation of 3D models, MyMiniFactory has a lot going on that you might like; in addition to letting you sign up to sell your models, the site hosts design competitions, often partnering with other companies in order to provide money prizes. And designers who want to get their name out there more—or who just love sharing what they know about 3D printing—can submit articles to the community blog.

To me, one of the most interesting features of MyMiniFactory is Scan the World, which describes itself as an "ambitious community-built initiative whose mission is to share 3D printable sculpture and cultural artifacts using democratized 3D scanning technologies." The project has partnered with museums and organizations around the world to make models of some of the most famous pieces in their collections and some of the world's most famous landmarks available for download and print. Fancy a replica of Michaelangelo's David to decorate your house? What about a miniature version of Florence's famed Duomo or the Great Mosque of Djenné in Mali to add a bit of pizzazz to a school report? How about playing chess with your very own set of the famed Lewis Chessmen? Scan the World states that its goal is "to bring tangible heritage to the masses," and I personally love the project. Check it out if you want to see some of the amazing things available for you to 3D print.

See all of this at www.myminifactory.com.

3D PRINTING SOFTWARE | 117

Etsy is one that might not have occurred to you, but you can find quite a few listings advertising STL files, along with designers offering their services for creating custom models. If you're looking for something specific and can't find it anywhere else, that's one place that might be worth checking out!

There are loads of other places you can check out; as you can see, the possibilities are nearly endless, and you may not find yourself ever needing to create your own 3D model! Get online and take a look at all the incredible models available to you.

PREPARE THE MODEL FOR PRINTING

So you've found a model you want to print: either you're using a test file, or you've downloaded an STL you found online to your computer. Now what?

You can't just print an STL file, as it turns out; it's basically just a list of coordinates describing the surface geometry of your 3D object, which is useless to a 3D printer. So you first need to prepare it in a piece of software called a slicer. What's a slicer, you ask?

Definition: *A slicer* is a piece of software that takes a 3D model and a series of user-entered settings and outputs a set of commands for a 3D printer.

See, we've talked a lot about how 3D printing works: the printer puts down layers that slowly build up until they've created the final object. The slicer, as you might guess from the name, slices up the model into those layers.

That's a simplification: what's happening is that you're inputting the 3D model, and you're also giving the software certain parameters, like layer height and printing speed. How you configure these parameters is going to vary depending on the model, your printer, the material, and your needs. All of these parameters interact with each other in ways that you need to be aware of. For instance, certain materials and certain printers will do better at certain speeds; certain models will work better with certain layer heights; and so on. Plus, you've got to think about the final purpose of the model and how much time you're willing to spend on it: a slower print speed and thinner layers might give you a more detailed final product, but it also could add hours or days onto your print time. And if all you're printing is a toothbrush cup for your RV, maybe you're not all that bothered about fine details, and you'd be happier with a quicker print. So as you can see, how these parameters are configured is going to vary from print to print, from printer to printer, from person to person.

Slicers generally allow you to make specific modifications as well. You can set the scale of the final product, making sure it's the right size you want. You can set how thick the walls of the printed object should be and whether parts are hollow or have an infill.

Definition: *Infill* refers to what's printed inside the walls of a 3D print. It can be of varying densities, patterns, and strengths. For instance, a printed object that needs to support weight or stand up to strain probably needs a higher density infill, while something that's just going to sit prettily on a shelf can probably get away with a low-density infill (recall that a lower density infill is going to require less print time, less material, and less money).

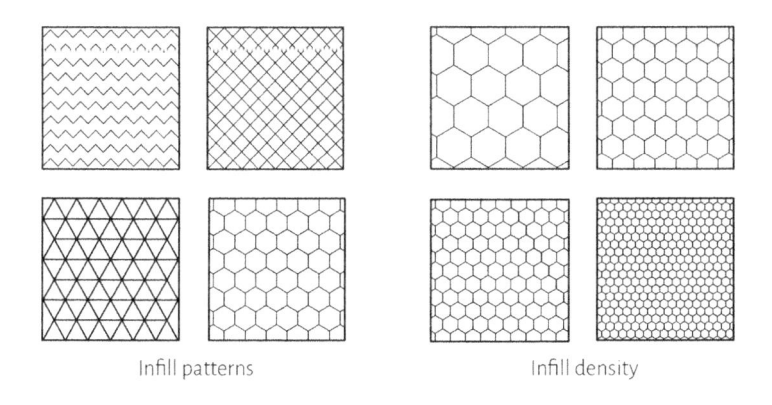

Infill patterns   Infill density

Finally, you can use the slicer to set up support structures. We've talked about this already; parts of the model that jut out or bridge over an open space may need support structures set up beneath them to keep them in place. Once the model is complete, you can remove them. The ability of a slicer to create support structures is a very important piece of functionality.

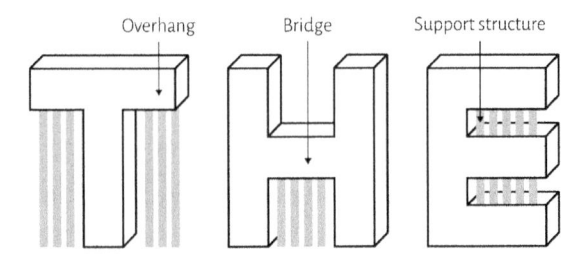

Support structures for overhangs and bridges

(Keep in mind, however, that not every bit that sticks out or bridges needs support structures. A good rule of thumb is that if each new layer in the overhang is less than 45 degrees, or if the bit that bridges are smaller than 5 mm, you probably don't need any support structures.)

Once you've added all of these inputs, the slicer will use all of this information to calculate the precise movements that the printhead needs to make in order to print the model. It then produces a set of commands to be sent to the printer. Generally, they do so using a language called G-code, which is also used in lots of other computer-aided manufacturing applications.

Getting a slicer

So how do you get your hands on a slicer? Well, if you don't want to put much effort into it, you're in luck: often, 3D printers come with a slicer, whether that's on some piece of storage media included with the package or available as a download from a website. The advantage to using this slicer (besides the ease of finding it) is that you know it's one that the printer manufacturer thinks will be suitable for your printer.

But maybe your printer doesn't have a recommended slicer. Or perhaps it does, but you don't think it's very good. Maybe you've been using it, and you think that another slicer with better controls (or even just different controls) could give you the results you've been looking for. In that case, you might look for a different slicer. Luckily, you can use nearly any slicer with nearly any printer; the G-code that a slicer outputs is pretty universal. (Still, just to be sure, before using a slicer you may want to Google it plus your printer model and see if anyone else has tried it before.)

When you do try a new slicer, it may take a bit of fiddling to get it dialed in to get the results you want. Remember that every combination of printer, material, model, and slicer is going to be a little different, and it may take a little doing to find the settings that will be just right for your particular printer, material, model, and slicer. Keep at it, keep adjusting settings, and consider doing a small test print before jumping into a massive model that's going to require a whole bunch of material.

There are lots of great slicers out there. The choice you make should be based on a few things: how much are you willing to pay? You can spend a fair bit of money on a great piece of software if you're so inclined. But there are also many great free slicers out there that may do more than enough for your needs.

Also, consider the features you need: a free option will have fewer cool features, while one you pay for will generally have a lot more. However, what you can get with a free slicer is often more than enough for your standard print jobs. Why pay for more features if you don't need them?

Generally, I recommend that you start with a free slicer, and later, when you've got a lot of experience under your belt, you can decide if you need something that can do more than your current software. After all, you can always spend more money when you decide you need to; you can't un-spend money if you realize you didn't need the fancy software.

Finally, you'll also need to check compatibility; generally, slicers and printers will have lists of the printers and slicers they're compatible with.

So with all that in mind, here are a couple of options you might consider when looking for a slicer:

Cura is generally agreed on as one of the best, if not the absolute best, free slicers out there. It's open-source, and if you know what you're doing, you can do quite a lot with it. It's pretty easy to use for beginners; for ease, you can just start out with its recommended settings and then branch out from there as needed.

But that simplicity of use doesn't translate into the simplicity of features; it can really do quite a lot, considering it's free. Its features include a preview stage, where it tries to identify potential failure points and the ability to monitor prints remotely.

And you know how we talked about how it's great to choose a popular printer because you'll find a lot of people online talking about it, and you'll be able to get help and useful information from that community? Cura's a lot like that; its popularity means you can find a lot of users, a lot of forums and websites and online groups, a lot of information, and a lot of help online.

Basically, if you want to move away from the slicer that came with your printer, I'd recommend starting with Cura.

Cura is made by Ultimaker, which manufactures 3D printers; fortunately, you don't need one of their printers to use the software.

Download it for free at ultimaker.com/software/ultimaker-cura.

If Cura is generally agreed on as the best free slicer, then **Simplify3D** is generally agreed on as the best paid-for slicer. If you want to get serious about your printing, this might be the way to go. Simplify3D's compatibility list is impressively long; they boast that they're compatible with more 3D printers than any other software out there.

It has a lot of great features, including realistic simulations of prints that let you see possible issues before you ever start printing (while Cura offers previews as well, Simplify3D's offering is, as you might guess, a cut above the Cura version). And the company offers something that you're probably not going to find with a free slicer: experts you can reach out to for help.

But of course, this doesn't come without a price: $149 USD for a license. However, they do have a two-week trial period, so you can try it and see if you like it before you commit to spending the money.

This is definitely one I'd recommend for more advanced users; it's probably not one to start out your 3D printing journey. But if you reach the point that you need those more advanced features, it's a great option.

Check it out at simplify3d.com.

While the two slicers above are probably the most widely used, I wanted to give a quick shout out to a couple of other options out there:

Tinkerine is a Canadian company that focuses specifically on 3D printers in educational settings: using 3D printers in classrooms to teach applied creativity. To that end, they have a very simple, easy-to-use slicer that is, intriguingly, entirely cloud-based. This isn't going to be a great option if you've got really complex models you'd like to prepare for printing, but if you're using your 3D printer in an educational setting—or even if you're just using it with a child—their easy-to-use interface might be perfect for your needs. And the fact that it's all done in your browser means that you don't need to download any software to your desktop. And it's free!

Check it out at tinkerine.com.

Slic3r is basically the complete opposite of Tinkerine: it has loads of great features, but (and this is often the curse of open source software) it's not necessarily the easiest interface in the world to use, especially if you're new to 3D printing. So why am I bringing it up? Because this is basically the grandfather of slicers! It's been around since the early days of 3D printing and has always been a free, open-source, non-profit endeavor. It's highly influential; Prusa based their own slicer off it. And there's a large community of users and developers surrounding it. Like Simplify3D, it's not necessarily the best slicer to start with, but as you dive deeper into the world of 3D printing, this is one you might want to check out.

Get it at slic3r.org.

SEND THE MODEL TO THE PRINTER

All right, so you've got your G-code ready for printing. The last thing to do is send it to the printer. How this all works will vary based on your printer and your slicer, so you'll want to read the instructions for each closely.

You also need to consider how to physically get the files from your computer to your printer. You have a couple of options: as with a desktop printer, you can hook it up with a USB cable, install some drivers (you may need to get on the website for the printer manufacturer to download the required drivers), and have your software send the files directly to the printer. However, with some printers, you have the option of putting the print job on an SD card and inserting that into a slot on the printer. Some people prefer this because you don't have to leave your computer plugged into the printer during long print jobs. (However, some printers download the entire print job at the beginning, so you don't have to leave the USB cable plugged in.)

If you want to plug it in, you'll likely need control software. Be aware that some slicers can act as control software for your printer, and some slicers require you to get separate control software. For instance, Slic3r functions only as a slicer, and you need another piece of software, such as Repetier or Repsnapper, to send it to the printer. Simplify3D, however, can talk directly to your printer.

Like I said, check the manual or search online to learn what your printer and slicer require.

And that's it! Now you have all the pieces—hardware and software—for doing your first 3D print job.

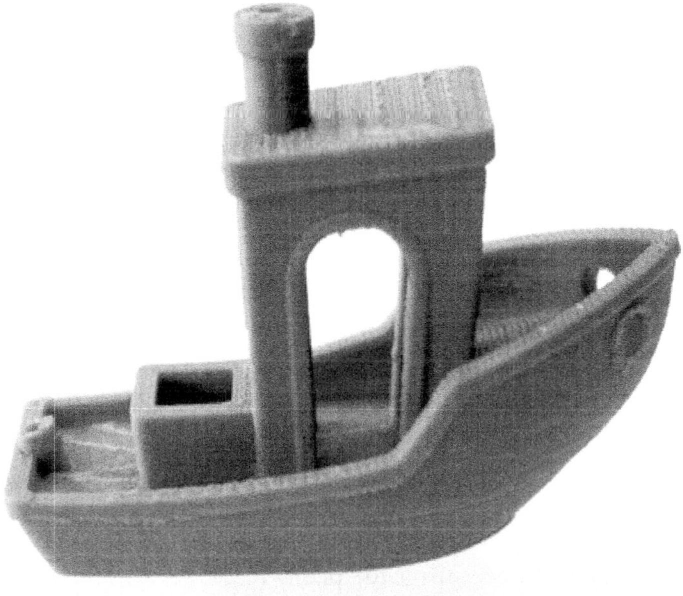

… truncated…

FIRST PRINT: EASY STEP-BY-STEP INSTRUCTIONS

So, now that we have the software and hardware ready; let's go over the steps for your first print.

1. Prepare your file

We just went over all of this in the last chapter, so I won't say much here: you've already heard about how to find an STL file and use a slicer to prepare it for print. If you're going to print from an SD card, get that SD card ready; if you're going to connect to the printer via cable, make sure that it's ready and the necessary drivers are installed.

One tip I will give you for your very first print is to choose a relatively small, simple print. You're still trying to get the hang of this, and you don't want to waste a bunch of filament if you get something wrong. Also, it's your first print! You probably want it to get done pretty quickly so you can see how it went rather than waiting nine hours for the printer to finish.

2. Prepare the print bed

There are three steps to preparing the print bed.

First, we need to **prepare the platform for proper adhesion**.

This has come up a few times already when we were talking about accessories and also about different types of filaments. You may remember that some filaments are going to do better at this than others. If you're using a material that is known for having too little or too much bed adhesion, you'll want to consider some of the following fixes in order to get the bottom layer to adhere to the print bed properly:

- We already mentioned the possibility of purchasing accessories to help.
- You can buy adhesives that you spread on the print bed to help it stick, like WolfBite or 3D Gloop.
- If removing the print at the end is your concern, there are some build plates that you can purchase that will flex, so when the print is done, you can lift the build plate off the print bed and bend it, and the print will pop off easier.
- Remember that certain print beds will interact with different materials in different ways. It's usually easier

to get a print off of a glass print bed, for instance. You can also buy print beds that are made of materials that are meant to help with adhesion: for instance, I've seen beds made of garolite advertised as being good with nylon prints, though I can't personally confirm that it works.

- Not interested in spending that kind of money? There are DIY options as well.
- If you're printing with ABS, consider spreading a glue stick—just a regular old glue stick, like you'd use to do a craft project in kindergarten—over the print bed. I'd recommend you do this only if you're using a glass bed. Just be careful that you're not letting too much glue build up in one place; the last thing we want is for the print bed to be lumpy.
- If you're printing with PLA, hairspray is a popular option; just spray a quick, even layer over the print bed.
- Another popular option is tape: just the blue painter's tape, the type you use to make sure you have clean lines when you're painting a wall. Just be sure to lay the tape down carefully: you don't want to leave any gaps between strips, but you also don't want any overlapping. Again, it's vital that the print bed is even. Some people also swear by Kapton tape, which is a heat-resistant tape with a shiny gold surface.
- If you're printing ABS, you can create a slurry made of

about 6 inches of ABS filament dissolved in about 2 ounces of acetone. The final liquid should be a little thicker than water, but not much. Once you're done, you can spread that over the bed (which should already be heated) with a brush. This is an effective solution but a little more complicated than glue sticks or tape.
- Remember that a heated print bed is designed to keep the lower layers evenly heated until the entire print is done; sometimes, that alone is enough to prevent any problems.

You may find that you need one, some, or none of these methods for proper bed adhesion. Maybe you'll have the best results from a heated bed that you've both taped and hairsprayed. But possibly, with an easy-to-use material like PLA and the right print bed, you don't need to do anything at all.

As always, before you use a bunch of filament on a massive print, try small test prints until you're sure you've got an option that's going to work for you.

Once that's ready, you need to **level the bed**. Since getting the print to work correctly has a lot to do with the Z-axis (the up and down movement of the printhead), you want to make sure an unlevel bed doesn't cause you any problems. If your print bed isn't level, it can mess up your bottom layers.

There are a few ways that you can do this:

- Printer leveling: Some printers will come with some kind of built-in method for leveling the bed. This may involve mapping out the heights of various points on the bed (using a sensor in the printhead) and changing the printhead's movement slightly to accommodate the slant. Processes like this tend to be easier but not always as accurate as you'd get from manual leveling.
- Software leveling: Some slicers and printer control programs, like Cura, have features to help you level your bed. See if the software you're using does, and follow the instructions to use it.
- Manual leveling: For the most accurate results, many people prefer to level the bed by hand. This definitely requires the most work, but it's a great thing to learn how to do if accuracy is important to you.

Here's how to manually level the bed.

1. Carefully clean the nozzle and the print bed before starting (unless this is your very first print).
2. Many printers have either three or four screws beneath the bed, carefully positioned in the corners (or two in corners and one in the opposite side's center, to form a triangle), which can be turned to lift that part of the bed. Locate the screws on your printer. (Certain

printers don't have these screws; if you have one of those, you're just out of luck here.)
3. If you're going to heat the print bed for this print job, some people recommend heating up the print bed now because the print bed can expand and contract as it's heated and cooled, so it's best to level it under the same circumstances that you mean to print. (This does mean the bed will be pretty hot, so if you do this, be careful!)
4. Move the printhead until it's right above one of the screws. Usually, you can do this just by moving the printhead, or the print bed, by hand.
5. Use the control screen on your printer if you have one, or the control software if you don't, to home the Z-axis —that is, lower the printhead to level 0. (If you're not sure how to do all this on your particular printer, check the manual.) At this point, there should be just a tiny bit of space between the nozzle and the print bed.
6. Get some paper: either a small piece cut from a piece of printer paper or an index card. Slide the paper between the nozzle and the print bed. You should be able to slide it in there, but you want there to be some resistance—just not quite enough to make the paper buckle or fold. If there's too much or too little space, use the screw to adjust the height at that point until it's where you want it to be.
7. Repeat with all other screws.
8. Once you've done all the screws, do them again, maybe

a few times. This will help you hone in on the leveling, plus every time you adjust one screw, you'll be affecting the others. It usually takes a couple of rounds of adjustments to get everything right.

Lastly, if you haven't done so already as part of the leveling process and need a heated bed for your print job, **preheat the print bed** using the printer control screen. Look into whatever temperature is recommended for the material you're using.

Now you're ready to print!

3. Get your filament ready

Now you need to make sure the correct filament is loaded on your printer.

Feeding the filament into the printhead can vary slightly based on whether your printhead includes a low-filament sensor; check your manual for precise instructions. But the basic idea is this:

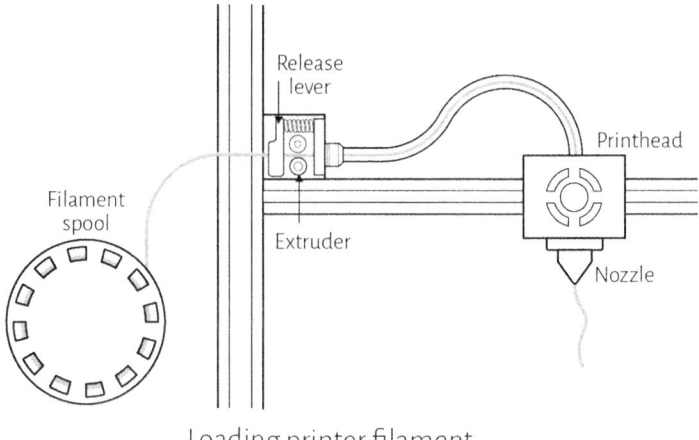

Loading printer filament

1. Use the printer control screen to start heating up the nozzle. Choose the temperature based on what's recommended for the material you're using; certain types need higher temperatures than others.
2. Clip the end of your filament at a 45-degree angle; this makes it easier to feed into all the places it needs to go.
3. Feed the filament through the low-filament sensor (if there is one).
4. Feed the filament through the extruder. Often, you'll need to push or squeeze a release lever to loosen up the extruder gears so you can pass the filament through them.
5. Generally, the extruder is connected to the nozzle via a tube. Keep feeding the filament into the extruder and through this tube until it reaches the nozzle.
6. If the nozzle is heated, then once the filament reaches

it, it'll begin to melt and come out of the nozzle. And you're done!

Always be careful with a spool of filament; don't let the filament loosen and tangle.

4. Print

Finally, it's time to print! If you're using an SD card, insert it now; generally, you'll use the printer's control screen to navigate to the file you want to print. If you're using a USB cable, you may just need to use your control software to tell the printer to start printing.

And finally, all your hard work is done! It's probably wise to keep an eye on the first few layers, just to be safe, but once you feel like the print job is off to a good start, you can sit back and let the printer do its thing. (Don't leave it entirely alone for long periods, though; since fires are a real, though not terribly common, possibility, it's best to stay reasonably nearby.)

5. Remove the finished print

It's minutes or hours later, and the print job is finally finished. You did it! You have your very first print, done with your own hands! It is time to yank it off the print bed and show it off to all your loved ones, right?

Wrong! There's a right and a wrong way to deal with a print once it's completed, and I bet you'd hate to have all your hard work ruined after the print is completed.

So, to avoid tripping at the finish line:

Wait

First, wait. Turn off the printer, or at least the heating of the print bed, and let the finished item sit. Personally, I wait until the whole print has returned to room temperature; I'd hate to warp it by prying it up when it's still got a soft, gooey center. For a small print, this can be a matter of minutes; for a large or dense print, it can take up to a couple of hours.

While you're waiting would be a great time to put your filament away. Pull the filament back out of the printer, carefully wind it back around the spool (you do not want your filament to get tangled, believe me), and put it back in wherever you're storing it to keep it tidy and dry.

Remove

Removing the print from the bed can be easy as pie, or it can be the hardest part of the print. But if you're careful, you can make it work.

If you used a flexible plate, just pick it up and flex it! If not, often, the print will just come up off the bed as it cools, if you're lucky.

If you're not lucky and it stays on the bed, you can try—gently!—to twist or pull at the print, holding very close to the bottom.

If this doesn't work, you may need to use tools to get it off: paint scrapers and spatulas are popular. The downsides to this method are that tools can damage the print bed (or yourself) if you're not careful. So be careful!

If you're using a scraper or spatula, don't try to scrape the whole thing off; you'll just damage the print bed and maybe the print. Try this instead:

1. Place the scraper or spatula at a spot where the print meets the bed; try to get the edge of it right at that spot where the two meet.
2. Use another object—something with a bit of weight but not too much, like the handle of a butter knife—to gently tap at the handle of the scraper or spatula. You may want to gently wiggle the scraper/spatula as well.
3. You may get lucky at that spot, and the print will pop up; if not, slide the scraper/spatula underneath it even more to work on getting it loose. If needed, move to another and keep going. Repeat until the print releases from the print bed.

Clean

Once the nozzle and print bed have cooled down, clean them as necessary. (Hopefully, you listened to my excellent advice and already put the filament away.)

I like to gently remove any material from the outside of the nozzle with a wire brush. If the nozzle has gotten clogged up, you may need to remove it from the printer and try to remove the clog manually (a needle might help here). You can also buy special cleaning filaments that are intended to remove blockages.

If there are any bits of material stuck to the print bed, remove them. You will likely not want to thoroughly clean the print bed after every print because a lot of the bed adhesion solutions we talked about can be used for multiple prints. If you do need to clean the print bed, using a lint-free cloth and rubbing alcohol may be a good choice; soap and water can also work, but I recommend that only if the print bed can be removed. You don't want to risk getting soapy water all over your printer.

Show off

Now you can take your final product to all your friends and amaze them that you created something from nothing. Well, from a spool of filament, but you know what I mean. You did it! You made your first 3D print. Now you can move on to bigger and better prints!

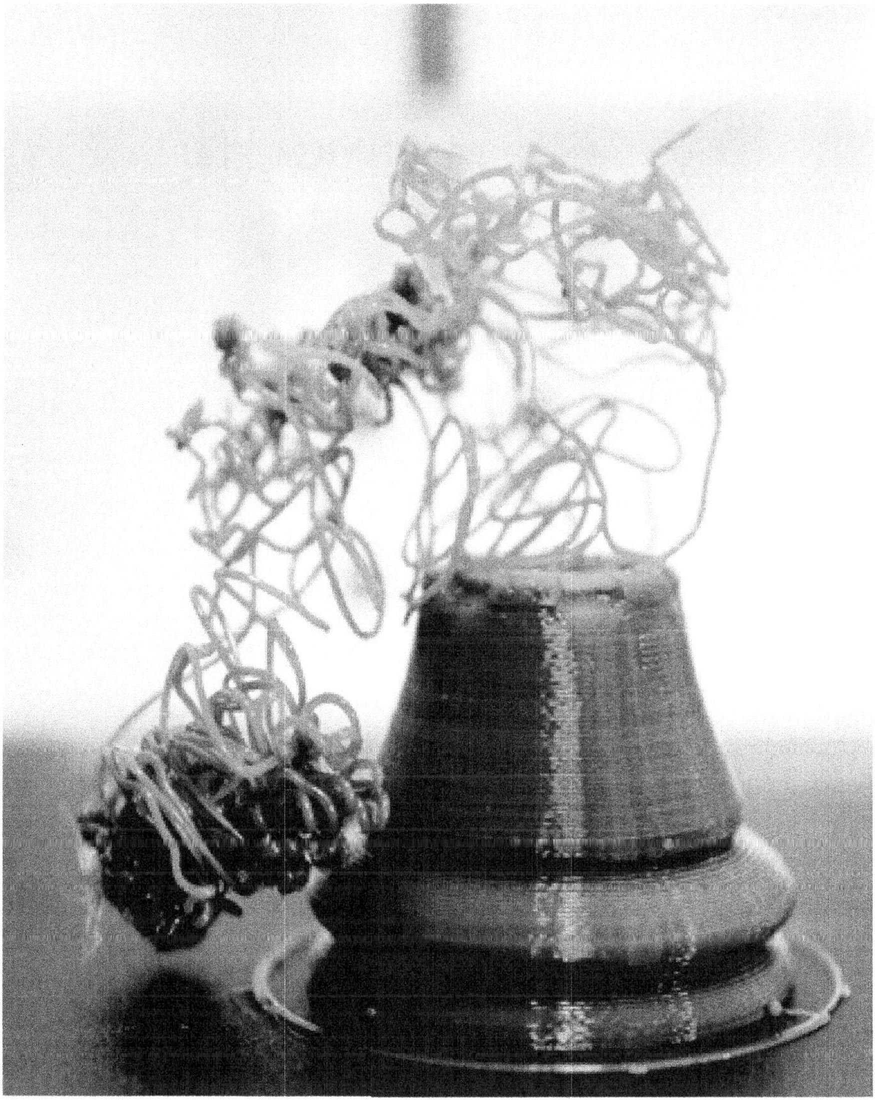

7

10 COMMON 3D PRINTING MISTAKES

All right, so you've done your first print; how did it go? Was the experience perfect the first time, or could it use some improvement? If the first: good for you! If the second, don't feel bad—it happens to all of us, even those of us who've been doing this for a long time. 3D printers are complex machines with lots of moving parts—both literal and figurative—which means there are lots of places where things can go wrong.

But just because it happens doesn't mean you just have to sit there and put up with it! There are lots of tweaks you can make to your process and your machinery to make sure your prints and your experience are good. We're going to talk about ten common mistakes that can mess up your 3D printing experience and how to avoid or fix them.

MISTAKE 1: NOT GETTING SERIOUS ABOUT THE FIRST LAYER

I put this first because it is by far the biggest mistake. The first layer is (literally) the foundation of everything that comes after it. If you don't get that right, the chances of the print working the way you want are not great.

So what do I mean when I talk about the first layer? I'm talking about bed adhesion, which we've discussed in previous chapters. Suppose you don't get the filament to stick the right amount to the print bed. In that case, you can see a host of problems: the object you're printing can warp (the edges of the first layer start to peel up, so you don't have a flat bottom to your print), or the object can get knocked out of place mid-print because it's not sufficiently stuck to the bed.

So what can we do about it? See the section on print bed accessories in chapter 3 for information about accessories you can purchase to help increase adhesion. And if you prefer a DIY approach, see the section called "Prepare the print bed" in chapter 6 for fixes you can do with items you can find around the house.

Here's another useful tip: do a quick test run on a small object before you do your real print. Then you can see if the object is going to stick before you commit to a large print; if it fails, is going to waste a lot of filament.

MISTAKE 2: NOT DOING TEST PRINTS

Speaking of test prints, I'm a big fan of them, particularly if you've changed some aspect of your print. If you're using the same material, the same print bed, the same temperature that you always do, then you definitely don't need to do this before every print. But if you've made a big change—this is a new material, you're trying a new temperature, you've got a new nozzle, you're trying a new print bed treatment—consider printing something small first.

Because like I said, 3D printers are complex, with lots of factors that affect them: the material, the print bed, the software settings, even the temperature of the room, if it's especially low or high. There are lots of things that could cause a print to fail. And wouldn't you rather have a small print that you don't care about fail, as opposed to a large print that is going to waste filament and money? If you really don't want to do a test print, at the very least, keep an eye on a print job if it's one of these where you've just made some major change to your setup. Hopefully, you'll be able to catch it if something starts going wrong.

While we're on the subject of test prints, there's another kind of test print that's useful: torture tests, which are models that are specifically designed to be difficult to print (they may involve overhangs, bridges, small details, curves, and more). They can be used to put a new printer through its paces or to calibrate a printer. You can find these all over Thingiverse; just search for "torture test." The most famous one is a little boat, which you've probably seen pictures of before if you've read a lot of reviews of 3D printers online.

MISTAKE 3: STORING FILAMENT IMPROPERLY

Your filament is the lifeblood of your 3D printer if you'll excuse the dramatics. You can have every other setting and piece of hardware dialed in perfectly, but if the filament—the actual material from which the print is made—is having trouble, your print is just not going to work out correctly.

There are a few common problems relating to a filament that you can run into, and they can all be fixed by careful storage and handling:

Letting filament absorb moisture

As I've mentioned previously, many of the filaments you will use for 3D printing can absorb moisture from the air (the word for this is "hygroscopic"); nylon and PVA are especially bad, but most filaments suffer from this problem to at least some extent. The danger with this is that if your filament absorbs a lot of moisture, then when it goes through the nozzle and heats up, the moisture will start escaping as steam (you may hear cracking or popping sounds as this happens). This can result in lower-quality prints in terms of surface appearance, layer adhesion, and material strength. It may also cause your nozzle and extruder assembly to clog.

So, what can be done about it? Proper storage is the first step. If you're really serious about this or if you live in a humid place, you might consider buying a special dry box to store the filament. At their most simple, these boxes are an airtight place to store filament, usually with something to dry the air in there, like a desiccant. Fancier options include boxes that heat up and dry the filament out, with built-in humidity sensors and everything. You can even get ones with spools and holes for the filament to be fed through to your printer, so if you want, your nice filament spools never have to meet the open air: they live in the storage container full-time. As you might guess, this sort of solution doesn't come cheap, but if you're ruining a lot of spools of an expensive type of filament, the investment might be worth it.

There are DIY options as well if that's more your thing (or if you're less worried about moisture, you live in a drier area, or you're just more frugal). Anything that seals up airtight can be used for storage—plastic bags, storage boxes—and you can purchase desiccant packages or use a heating element, such as those used in reptile cages, to keep the filament dry. And if you want a fancy box that lets you feed the filament to the printer without taking the spool out, there are tutorials online to teach you how to construct them for far cheaper than what you would pay for one online.

Letting filament get dusty

A similar concern is dust: as anyone who's tried to keep a room clean can tell you, anything that sits out will get dusty eventually. Dust on your filament can cause the extruder and nozzle to get clogged.

As with the moisture issue, a storage box can take care of that for you. Another option to consider is a filament filter, which you can conveniently 3D print and assemble at home. This is a tiny part, usually cylindrical, with a bit of sponge moistened with mineral oil inside. If you thread your filament through this filter before inserting it into the extruder, the filament will be cleaned of dust as it gets pulled through the filter. You can find a number of different models for filters online.

Letting filament get tangled

Filament on a spool can loosen if the cut end of the filament is allowed to run free, and then it can tangle. And if your filament gets so tangled that the spool no longer spins freely, that could cause a serious problem while you print.

Fortunately, most filament manufacturers are very conscientious about not giving you tangled spools. Unfortunately, that means that most tangled spools are caused by user error: either careless storage or careless handling of filament spools. Generally, if you let the filament loosen up and you aren't careful when you retighten it, the filament can cross over itself, causing a tangle.

There are two major ways to prevent this:

- First, when handling a spool, always keep the end of the filament in one hand and pull it taut enough that the filament on the spool can't loosen.
- Second, when storing the filament, don't allow the end of the filament to hang loosely; you can clip or tape it down, or with some spools, pass the end of the filament through the holes in the side of the spool until it stays.

The most important thing here is that the end of the filament is always secured and pulled tight enough that the rest of the filament isn't going to loosen.

If you do end up with a tangle in the spool, carefully unwind the filament until you find the tangle (you may have better luck pulling loops of the filament over the edge of the spool, one loop at a time). Then carefully re-wind it, laying each loop down side by side.

That's an arduous process, though; I'd recommend you do your best not to get a tangle in the first place.

MISTAKE 4: NOT PREPPING THE MODEL PROPERLY

I can't overstate the importance of getting the slicer step right. This is where you turn a model into a reality, into something the 3D printer can actually print. So take this bit seriously! A few things to keep an eye on:

- Figure out what settings are best for your printer and your material. This is where a bit of a test print could come in handy.
- Use support structures. The slicer is where you will be able to add support structures to the model. We've talked about this at length already, but in case you skipped to this page first: you need support structures to prop up bits of the model that extend out into space or that bridge over a gap. When the print is done, you remove the structures (often by cutting them off). If you don't have that support under overhangs and bridges, there will be no lower layers for the upper layers to be printed on, and the bottoms of those overhangs and bridges might look like a messy plate of spaghetti.
- Rotate the model. Not that keen on using too many support structures? You may be able to avoid a lot of difficulties in your print by simply rotating. Imagine

printing a capital letter T: if it's standing upright, you'll need support structures under the crosspiece. But if it's lying down on its back, you don't need any support structures at all.

MISTAKE 5: NOT LEVELING THE BED

So, we talked a lot about leveling the print bed in the previous chapter, but here's the gist: having your bed as level as possible is important for a good quality print. Under normal operation, your printer will assume the print bed is level and calculate the height of the printhead accordingly. If the bed isn't level, you may have places in your print where you don't get good adhesion on the bottom layer because the print bed was too far away when the printhead laid down that layer.

If you want more details about how to go about this process, see the previous chapter.

MISTAKE 6: IGNORING THE PRINTER WHILE IT'S IN USE

The number one reason you'll hear from people about why not to leave a print job completely unattended is safety. There have been cases of 3D printers starting fires, and if that's going to happen, you want it to happen when you're monitoring the print job so you can respond in time. Now, because this is a known problem, printers these days are often equipped with what's called thermal runaway protection, where they will turn off if there's a fault or other undesirable conditions occur. But even that isn't a guarantee that nothing will ever go wrong.

Another reason it's a good idea to monitor prints is so that you can stop the print if it fails. If something goes wrong with the print or if the object you're printing gets knocked over, you don't want the printer to carry on to the end of the print job—what a waste of filament! If you keep an eye on the printer, you can stop it if you need to.

"So what?" you're saying. "You mean I need to sit in the room and watch the printer for all fourteen hours of a print?" Of course not; that would be a waste of your time. But there are a few things you can do; here's what I do recommend:

- Use a printer with thermal runaway protection, if you can; it's not a fail-proof solution, but it's certainly better than nothing.
- Make sure all parts of the printer are in good repair.
- Check-in on the printer occasionally during a print job. One good rule of thumb is to watch the first couple layers to make sure there's not an immediate problem, and once you're satisfied that it's off to a good start, come back and check on it every so often—every half-hour to an hour, perhaps.
- Some printers come with a camera that allows you to monitor a print job remotely; the Flashforge

Adventurer 3 is an example of one such printer. With one of these, your spot checks can be done remotely, from a phone or computer.

- For your first few print jobs with a new printer, I'd recommend that you don't leave it unattended for long, just in case there's been an error in your assembly or one of the parts is faulty. Once you've used it a few times and are feeling a little more confident, you don't need to babysit it as much, though I still recommend you don't start a new print job and then take off for a weekend in Cabo.

MISTAKE 7: NOT TAKING SAFETY PRECAUTIONS

The fact that consumer 3D printers are accessible to the average person is part of their appeal. But don't fall into the trap of thinking this means that 3D printers are just like the laserjet printer that's sat on your computer desk since 1996. These 3D printers are great, but they can definitely cause more problems than your standard desktop printer.

The first thing you'll want to do is **make sure the printer area is properly ventilated**, as certain filaments can release strong odors—and even become dangerous. ABS and ASA are known for having a problem here, but other filaments can have problems as well. It's good to get in the habit of ventilating the print area, even when you're not using a filament that's known to be dangerous.

Also, **watch out for hot surfaces.** The print bed and nozzle can get up to hundreds of degrees, and if you're not careful, you can burn yourself before, during, or after a print (remember that after the print is done and the printer is off, it can take a while for the metal surfaces to cool down).

Speaking of hot things, **beware of fires.** We talked about this above, but you'll want to take precautions where possible and keep an eye on the printer. It also helps if you make sure the printer is in good repair and replace parts as necessary.

Lastly, **be careful when removing prints.** Many people find that using a tool like a spatula is best for removing prints that are stuck to the print bed, but I've heard multiple stories of people injuring themselves while using these tools. And a slipping spatula or screwdriver can hurt not only your hands but also the surface of the print bed. So be thoughtful and cautious any time you use these tools.

MISTAKE 8: IGNORING MAINTENANCE

As with any machine, the parts on a 3D printer can wear out or become damaged over time; you can't just build the printer once and then hope for the best forever. And even when parts aren't getting damaged, they can get loosened over time.

These kinds of problems can manifest themselves in lots of ways. A damaged (or clogged) nozzle can cause prints to be of lower quality. Loose belts can cause pieces not to move as they should, leading to prints failing. And as we mentioned above, some hardware problems can lead to fires.

So keep an eye on this! Watch for any degradation in your print quality. And check your printer often, looking for loose nuts and bolts. This will help you to have high-quality prints and to avoid sticky situations like fires.

MISTAKE 9: TRYING TO DO IT ALL YOURSELF

Let me tell you about me: I am terrible at asking for help. When I get a new gadget or appliance, or if I need to fix one I already have, I first waste some time trying to figure it out myself, and only when hard experience has proven that it isn't going to work do I reach out for help. Sound familiar to you? I know I'm not the only one who's like this.

Now, you've gotten this far in a book designed to help you get started with 3D printing, so clearly, you're at least a little bit willing to look for help. To you, I say: keep at it. Keep looking for help. There are certain things that you probably can figure out on your own, like programming a microwave. But this isn't programming a microwave. We're talking about a 3D printer, which is a complex piece of machinery controlled by a complex piece of software. The chances of getting everything—all the settings in both hardware and software—correct, just based on you trying to figure it out on your own, are not high.

And you don't want to mess up your prints, right? After all, your 3D printer and your filament cost you money.

So, as I said, get help. Start by reading the manual—after all, a technical writer somewhere worked hard to create that information for you—and by reading information like this very book.

Then get online and start looking for help. Find a group somewhere—Facebook, forums, etc.—that is full of people who use your very printer so that you can ask them for help. Find website articles that are full of useful information, and if you're a more visual learner, check out YouTube. There's a huge amount of absolutely amazing and helpful information available on 3D printing right now.

Learn from those who've gone before you; you don't have to learn by experimentation and failure and ruining multiple prints until you dial in on the perfect settings. Learn from other people's mistakes, and make the messy beginning of your 3D printing career easier.

MISTAKE 10: GIVING UP TOO EASILY

Let's circle back around to the first chapter of this book. My advice there still holds: keep realistic expectations and just keep trying. 3D printing isn't always easy. So if your first print or two or five doesn't work out the way you want, don't give up! Keep trying. When a print fails, fiddle with some settings, do some reading online, ask questions in a forum if needed, and try again.

You got this.

CONCLUSION

So here we are, at the end of the book! We've covered:

- What 3D printing is
- Picking a printer
- Helpful accessories
- 3D printing materials
- What software you'll need
- All the steps for your first print
- Ten common 3D printing mistakes and what to do about them.

I hope you've found this helpful, and I hope it's given you a good overview of the essential tips and tricks you need to get started on your 3D printing journey. 3D printing is an incredible hobby that has provided people all over the world with entertainment, a chance to exercise their creativity and improve their technical skills, and a way to easily fabricate objects that are useful, decorative, or just fun!

I hope this book has convinced you to join that number—and I hope it's convinced you that you can join that number! I know that it can seem a little daunting at first: so many decisions to make, so many settings to fiddle with. But the very fact that there are so many options is one of the things I love about 3D printing: it's so customizable. You can choose the printer that's right for you, with the accessories and materials right for you, plus the software that's right for you. You can customize the experience to what you want, what you're willing to pay for, how much time you're willing to spend learning . . . And then you can create exactly what you want: the object you want in the material you want in the size you want.

It does take some time and effort to learn the ropes, but don't forget that there's a whole world of resources (including this book) and people out there that want to help; you just have to be willing to look and occasionally to ask. Be confident (but humble enough to ask for help); be flexible if things don't go the way you want the first time. I know that you'll be printing like a pro in no time.

If you've liked the book, I'd appreciate you taking the time to leave a review on Amazon! And if you get farther into your journey and want a little more help, check out the later books in our series.

But for now, get out there and get printing!

LEAVE A REVIEW

Customer reviews

★★★★★ 5 out of 5

2 global ratings

5 star	100%
4 star	0%
3 star	0%
2 star	0%
1 star	0%

˅ How are ratings calculated?

Review this product

Share your thoughts with other customers

Write a customer review ⬅

I would be incredibly thankful if you could take just 60 seconds to write a brief review on Amazon, even if it's just a few sentences! You can do so by clicking the link below.
I love hearing from my readers, and I personally read every single review.

Click here to leave a quick review.

A SPECIAL GIFT TO MY READERS

Included with your purchase of this book is the *3D Printing: 10 Beliefs to Stop Now* list. This printable will reveal the most common and surprising myths about 3D printing.

Click the link below and let me know which email address to deliver it to.

nataliebooks.com

REFERENCES

A. (2017a, November 7). Understanding the ABS Plastic in LEGO. LEGO Ways. https://legoways.com/abs-plastic-in-lego/

C. (2017b, July 29). Are Dual Extruders Worth It? To Buy a 3D Printer. https://tobuya3dprinter.com/dual-extruders-worth/

Griffith, B. H. (2014, March 12). Pioneering 3D printing reshapes patient's face in Wales. BBC News. https://www.bbc.com/news/uk-wales-26534408

Lim, A. (2018, May 2). Could Bioprinting Save Your Life? ThoughtCo. https://www.thoughtco.com/what-is-bioprinting-4163337#:%7E:text=Bioprinting,%20a%20type%20of%203D%20printing%20,%20uses,organs,%20cells,%20and%20tissues%20in%20the%20human%20body.

McCue, T. J. (2020, March 4). The 5 Best Ways You Can Make Money With a 3D Printer. Lifewire. https://www.lifewire.com/make-money-with-a-3d-printer-2216

Peters, A. (2020, March 6). This village for the homeless just got a new addition: 3D-printed houses. Fast Company. https://www.fastcompany.com/90469488/this-village-for-the-homeless-just-got-a-new-addition-3d-printed-houses

Sharma, R. (2013, September 12). Custom Eyewear: The Next Focal Point For 3D Printing? Forbes. https://www.forbes.com/sites/rakeshsharma/2013/09/10/custom-eyewear-the-next-focal-point-for-3d-printing/

Speeney, R. (2020, March 31). Blue Tape for 3D Printing: The Complete Guide [2021]. TapeManBlue. https://tapemanblue.com/blogs/tips-tricks/blue-tape-for-3d-printing

Varnak. (2020, December 30). Why Your Next 3D Printer Should Use a 32 Bit Controller. Mechlounge. https://mechlounge.com/why-your-next-3d-printer-should-use-a-32-bit-controller/

Printed in Great Britain
by Amazon